The Mackenzie

RIVERS OF AMERICA BOOKS

already published are:

KENNEBEC by Robert P. Tristram Coffin
UPPER MISSISSIPPI (New Revised Edition, 1944) by Walter Havighurst
SUWANNEE RIVER by Cecile Hulse Matschat
POWDER RIVER by Struthers Burt
THE JAMES (New Revised Edition, 1945) by Blair Niles
THE HUDSON by Carl Carmer
THE SACRAMENTO by Julian Dana
THE WABASH by William E. Wilson
THE ARKANSAS by Clyde Brion Davis
THE DELAWARE by Harry Emerson Wildes
THE ILLINOIS by James Gray
THE KAW by Floyd Benjamin Streeter
THE BRANDYWINE by Henry Seidel Canby
THE CHARLES by Arthur Bernon Tourtellot
THE KENTUCKY by T. D. Clark
THE SANGAMON by Edgar Lee Masters
THE ALLEGHENY by Frederick Way, Jr.

RIVERS OF AMERICA BOOKS

continued

THE WISCONSIN by August Derleth
LOWER MISSISSIPPI by Hodding Carter
THE ST. LAWRENCE by Henry Beston
THE CHICAGO by Harry Hansen
TWIN RIVERS: *The Raritan and the Passaic,* by Harry Emerson Wildes
THE HUMBOLDT by Dale L. Morgan
THE ST. JOHNS by Branch Cabell and A. J. Hanna
RIVERS OF THE EASTERN SHORE by Hulbert Footner
THE MISSOURI by Stanley Vestal
THE SALINAS by Anne B. Fisher
THE SHENANDOAH by Julia Davis
THE HOUSATONIC by Chard Powers Smith
THE COLORADO by Frank Waters
THE TENNESSEE: THE OLD RIVER, *Frontier to Secession,* by Donald Davidson
THE CONNECTICUT by Walter Hard
THE EVERGLADES by Marjory Stoneman Douglas
THE TENNESSEE: THE NEW RIVER, *Civil War to TVA,* by Donald Davidson
THE CHAGRES by John Easter Minter
THE MOHAWK by Codman Hislop
THE MACKENZIE by Leslie Roberts

SONGS OF THE RIVERS OF AMERICA, edited by Carl Carmer

Rivers of America

EDITED BY

HERVEY ALLEN *and* CARL CARMER

AS PLANNED AND STARTED BY

CONSTANCE LINDSAY SKINNER

Associate Editor

JEAN CRAWFORD

Art Editor

FAITH BALL

The Mackenzie

BY

LESLIE ROBERTS

ILLUSTRATED BY

THOREAU MACDONALD

GREENWOOD PRESS, PUBLISHERS
WESTPORT, CONNECTICUT

Library of Congress Cataloging in Publication Data

Roberts, Leslie, 1896-
The Mackenzie.

Reprint of the ed. published by Rinehart, New York, in series: Rivers of America.
Bibliography: p.
1. Mackenzie River. 2. Northwest Territories, Can. --History. I. Title. II. Series: Rivers of America.
F1100.M3R62 1974 917.193 73-20906
ISBN 0-8371-5864-8

Designed by Stefan Salter

Originally published in 1949 by Rinehart & Company, Inc., New York

Reprinted by arrangement with Holt, Rinehart & Winston, Inc.

Reprinted in 1974 by Greenwood Press,
a division of Williamhouse-Regency Inc.

Library of Congress Catalog Card Number 73-20906

ISBN 0-8371-5864-8

Printed in the United States of America

TO GLADYS

Contents

Foreword		xi
ONE	Highway to the Top of the World	3
TWO	Before the White Man Came	15
THREE	Alexander Mackenzie Finds a River	29
FOUR	Northwest Passage	49
FIVE	The Fur Barons	86
SIX	The Golden Years	113
SEVEN	A Free Country	143
EIGHT	The New Empire Builders	166
NINE	New Northwest Passage	209
TEN	The Road to the Top	242
	Bibliography	255
	Index	257

Foreword

THIS BOOK has been almost fifteen years in the making. When its author first saw the Mackenzie, in the early days of flight into its vast empire, he had no idea that he was beginning a book about one of the great Rivers of America. He went back and saw the country and, to show you what kind of country it is, he once flew more than three hundred miles on a winter's Saturday afternoon, for no better reason than to play stud poker with a Mountie, a fur trader, a mine manager, a geologist, and his own companion, an airplane pilot. Then, come Sunday, they, pilot and writer, flew back whence they came, to resume the chores that had brought them into the country.

Shortly after V-J day a third journey was made to the Mackenzie, in order to revisit the mines at Great Bear Lake where the makings of U-235 and The Bomb had been extracted from the frozen ground. It was on this visit, while flying over the great river, heading south from Canol and the oil wells at Norman, that the idea of this book took root, resulting in a letter to Stanley M. Rinehart, Jr., publisher of the Rivers of America series, suggesting that it was about time somebody wrote a book about this fabulous river and its basin.

Many people have been of immeasurable assistance, some without knowing it. This is to thank them, companions of the aerial argosy that finally cracked open Alexander Mackenzie's north: Leigh Brintnell, Wop May, Punch Dickins, Matt Berry, and Gilbert LaBine, all of whom will be met as you read. Such men not only made this book possible by being its author's friends and traveling companions; they and their kind made the new Northwest.

Finally, a writer's sincere thanks to three persons whose aid has been beyond measure: Mrs. Douglas Mackay, of Winnipeg, has been of great help in assembling data on the early days of aviation in the Mackenzie country; Mr. Hereward Senior, of Montreal, post-graduate student at McGill University, and my son, Mr. Grant Roberts, have been of incalculable assistance in piecing together the history of the fur trade and the voyages and overland journeys of that gallant band of men who came this way to seek the Northwest Passage.

Montreal, Canada
April 2, 1948

The Mackenzie

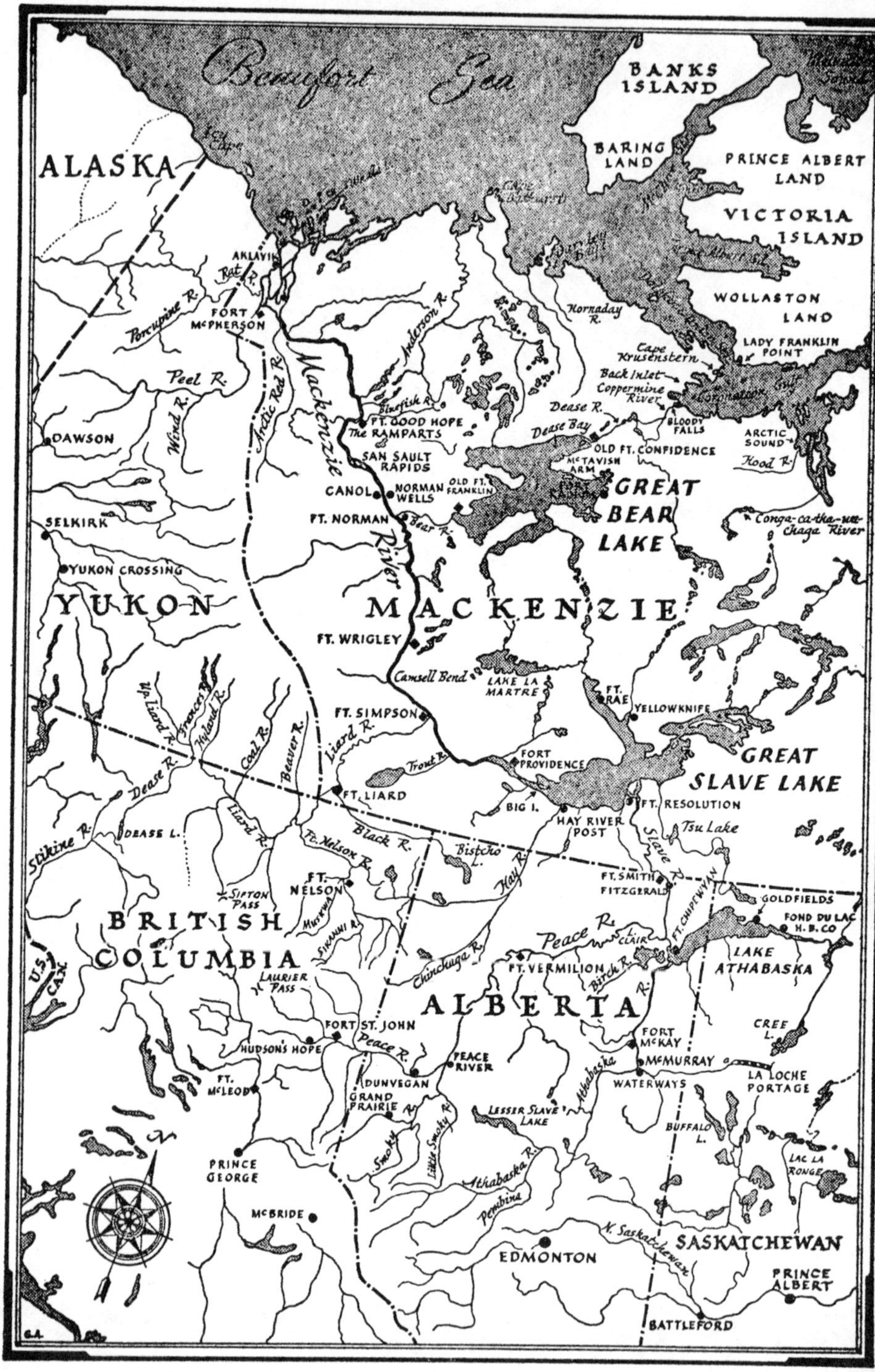

Beaufort Sea
BANKS ISLAND
ALASKA
BARING LAND
PRINCE ALBERT LAND
VICTORIA ISLAND
AKLAVIK
Rat R.
Porcupine R.
FORT McPHERSON
WOLLASTON LAND
Hornaday R.
Anderson R.
LADY FRANKLIN POINT
Cape Krusensbern
Back Inlet
Coppermine River
Peel R.
Wind R.
Arctic Red R.
Mackenzie River
Bluefish R.
Dease R.
Dease Bay
BLOODY FALLS
ARCTIC SOUND
Hood R.
DAWSON
FT. GOOD HOPE
The RAMPARTS
SAN SAULT RAPIDS
OLD FT. CONFIDENCE
McTAVISH ARM
CANOL
NORMAN WELLS
OLD FT. FRANKLIN
GREAT BEAR LAKE
SELKIRK
FT. NORMAN
Bear R.
Conga-ca-tha-wa-Chaga River
YUKON CROSSING
YUKON
MACKENZIE
FT. WRIGLEY
Camsell Bend
LAKE LA MARTRE
FT. RAE
YELLOWKNIFE
Up. Liard R.
Frances R.
Hyland R.
Coal R.
Beaver R.
FT. SIMPSON
Liard R.
Trout R.
FORT PROVIDENCE
GREAT SLAVE LAKE
Dease R.
FT. LIARD
BIG I.
FT. RESOLUTION
Stikine R.
DEASE L.
Liard R.
Ft. Nelson R.
Black R.
Bistcho L.
HAY RIVER POST
Tsu Lake
Slave R.
Hay R.
Sifton Pass
FT. NELSON
FT. SMITH
FITZGERALD
GOLDFIELDS
BRITISH COLUMBIA
Muskwa R.
Sikanni R.
Peace R.
L. CLAIR
FT. CHIPEWYAN
FOND DU LAC H.B.CO
LAKE ATHABASKA
U.S. CAN.
Chinchaga R.
FT. VERMILION
Birch R.
Laurier Pass
ALBERTA
FORT ST. JOHN
FORT McKAY
CREE L.
HUDSON'S HOPE
Peace R.
PEACE RIVER
McMURRAY
LA LOCHE PORTAGE
Athabaska
FT. McLEOD
DUNVEGAN
WATERWAYS
GRAND PRAIRIE
LESSER SLAVE LAKE
BUFFALO L.
Smoky R.
Little Smoky R.
PRINCE GEORGE
Athabaska R.
LAC LA RONGE
McBRIDE
Pembina
N. Saskatchewan
SASKATCHEWAN
EDMONTON
PRINCE ALBERT
BATTLEFORD

CHAPTER ONE

Highway to the Top of the World

ONE VITAL FACT will always make the Mackenzie distinctive among the great rivers of America. Each of the others in its day, Mississippi, Missouri, the St. Lawrence, has been a highway over which men have opened new frontiers until, one by one, the frontiers disappeared and the rivers became the traveled paths of our busy inland commerce. Not so the Mackenzie.

When the dour Scots trader from whom it takes its name first ventured out of Great Slave Lake in 1789 and turned north toward the Arctic Ocean, the river was already the road to trade and war between the Slaveys, who lived along its southern reaches, and the Hare Indians, whose lodges were built astride the Arctic Circle. But to the American interior's come-lately white masters, such few as had heard tell of it, it was a silver streak flowing north to Nowhere.

So it is to this day. Tens of millions of North Americans have seen and traveled the Mississippi and the St. Lawrence. Every Canadian and American schoolboy knows where to find the Hudson on the map. But not one per cent of one per cent of the continent's population have ever seen the great river that snakes through the heart of the million square miles of tundra that are Canada's Northwest Territories and not many could lay an index finger on its sinuous ribbon in the atlas. When the barges laden with uranium concentrates, which placed in America's hands the most lethal weapon ever devised by man, moved south over its waters, midway through World War II, and when the Army of the United States thrust a pipeline from its banks through the mountains to Alaskan tidewater, the Mackenzie was still the highway of the frontier.

So it will remain. When Aklavik in its broad arctic delta is a flourishing city, when Fort Norman is the center of a great oilfield and Yellowknife on Great Slave Lake has become, as it may, the richest gold camp in the lurid history of the gleaming metal, the Mackenzie and the lakes and rivers of its tributary watershed will still be the highway of the frontier, for frontier this will remain as long as mankind shall endure.

The Mackenzie is more than a river. It is the huge funnel

through which flow the accumulated waters of rivers greater than the Hudson and of lakes greater than Erie or Ontario. Thus, to restrict examination to the river alone would be to ignore the great sweep of a majestic canvas. Water flowing down to the northern sea through the valley of the Mackenzie has laved the fertile wheatlands of the Peace River country, carried machinery to the miners of Yellowknife and washed the radium-bearing rocks of Great Bear Lake. Even in the navigable sense it is a series of waters which make one. As the gentle Athabaska it flows north from Lesser Slave Lake, past the end of steel at Waterways, Alberta, whence shallow-draft steamers and barges move north during the brief ice-free season laden with the year's supply for trading posts, mining camps and oilfields. One hundred sixty-five miles north of the railroad tracks, after adding the waters of Lakes Athabaska and Clair to its own, it merges with the Peace and becomes the Slave, still the placid giant, until, after another 70 miles, it enters the turbulent seven miles of water that run between Fitzgerald and Fort Smith, the only unnavigable stretch in the 1,300 crowflight miles – and God knows how many hundreds more as the river twists and turns – which intervene between the end of the railroad and the northern sea.

A highway cuts across the carrying place today, over which trucks roar as they transship cargoes from barge to barge over the portage. A wagon track crossed the carrying place even before Gilbert LaBine found radium- and uranium-bearing rocks on the Arctic Circle, almost 500 miles to the north, early in the 1930's and set in motion the Territories' first wild mineral hunt. From that time forward the old fur traders' portage came to life with a whoop each season as soon as the ice left the Slave and the barges crowded close on its heels as the miners rushed machinery, explosives and foodstuffs north to see them through the next interminable subarctic winter.

But if the first years of the gold, radium and silver rush were years to astound — and to annoy — the slow-moving fur traders who, until the minemakers came, had held undisputed dominion over the whole Mackenzie watershed, they were years of the locust by comparison with what the impact of global war was to make happen down north. (Because the Mackenzie flows north, the usual terms "down south" and "up north" will be reversed.) Over those seven short miles of portage passed hundreds of thousands of tons of equipment and supply for the strange Canol adventure, the thrusting of an oil pipeline from the wells on the lower Mackenzie through the mountains to Pacific tidewater. Over the road passed tens of thousands of tons more in mining machinery and supply for the hurriedly reopened Eldorado radium-uranium mine on Great Bear. And over it, southbound, passed the sacks of coarse black gravel, concentrated pitchblende, heading for its final destination over Hiroshima. Through this one short bottleneck, then, passes the flood of traffic of a hinterland empire one million square miles in extent, an empire of which the Mackenzie and its tributary waters are the arterial system.

From Fort Smith, administrative capital of the Territories, with its Mounted Police barracks, mining recorder's office, Indian agent and hospital, navigable water stretches clear ahead, more than 1,200 miles to the Beaufort Sea. Until Great Slave Lake is reached, it is still the Slave River, winding so tortuously through its path between low, rocky hills that at times, seen from the air, it almost threatens to come full circle and join itself flowing in the opposite direction. As the plane flies, only 100 miles intervene between Smith and the lake. But as the river flows, almost twice that distance is traveled to complete the next leg of the arctic-bound journey.

The Great Slave crossing, more than a hundred miles of steaming due west across a lake larger than Erie and its equal in sudden treachery, is tricky going and the wise captain of delta-bound Hudson Bay packet, or of tug and barges, will lie-to in the lee of the river mouth, unless the barometer is set fair. Not even these precautions add up to full insurance, for the big lake can lash into a fury quick as a tugboat skipper can spit out his chaw, as mute evidence to which are numerous wrecks which strew the southern shores of Great Slave between Fort Resolution and the entrance into the Mackenzie proper. The impelling sense of haste that derives from the short open season, coupled with the trickiness of Great Slave weather, poses a constant conundrum to the down-north navigator, the urgency of which increases as the days of open water decrease.

Once across the big lake, however, a smooth journey

lies ahead for ships that were built for the rivers, not for the rough-and-tumble waters of the inland seas. This is the country of the Dogribs and Slaveys, of trappers and furs, where the river flows wide between low banks, past Fort Providence and on more than a hundred miles to Fort Simpson, where the Liard comes tumbling down from the western mountains.

Beyond Simpson the river drives northwest toward the great Camsell Bend, the rugged profile of the Mackenzie Range lining the broad valley in the distance. Sharply it turns north again, to flow gently on, 100 miles to Wrigley, whence it holds its northern bearing on almost undeviating course for another 150 miles to its junction with the fast-flowing waters of the Great Bear River, hard by Fort Norman fifty miles south of the abandoned Canol Project.

Around the rapids of the Bear, until recent days, roustabouts manhandled the cargoes of supply for the radium-uranium mines on Great Bear Lake and, southbound, packed out the heavy sacks of concentrate. Now a Canadian government road has been thrust across the muskeg (the goo and gumbo swamp country composed of rotting, fallen timber, rocks and casual water which abounds in low-lying country in northen Canada) from the foot of the rapids to the site of old Fort Franklin in the southwest corner of Great Bear Lake, third in size of North America's great freshwater seas. From the end of the highway on the shore of the jag-coasted lake astraddle the Arctic Circle cargoes are transshipped again, this time across two hundred miles of open water to what was the hemisphere's most hush-hush wartime minesite, the pitchblende workings at Port Radium, up to now North America's only major domestic source of fissionable material.

After the Yanks pulled out in 1945, leaving their pipeline behind them, the left bank of the Mackenzie at Norman Wells could justifiably be termed the junkyard of the subarctic. Across the huge flats around the Canol base the departing engineers strewed bulldozers, jeeps, all manner of mechanized military equipment, abandoned by its American owners because to lug it south over the tortuous waterway would cost Uncle Sam far more than any price it might bring in the open market. So ended an adventure which cost the American people more than $400,000,000.

From the capped oil wells of Norman the Mackenzie flows on through what is still the unchallenged fur empire. Even trading posts are few and far between across the open wilderness. Here and there on the riverbank stands a rough-hewn trapper's cabin, here and there a minuscule Indian settlement. Past the ramparts of Fort Good Hope and the hutments of Arctic Red River, the Mackenzie rolls slowly along until, more than a hundred miles before emptying into the Beaufort Sea, it begins to divide and redivide into the series of mouths that carry it through the huge delta into salt water. On the main western channel stands Aklavik, a neat and tidy, white-painted settlement with government hospital, police barracks, and other signs of the law north of the Circle, which are its only valid reason for existence. Thence the river flows on, divided and alone, through the treeless emptiness, until it loses itself in the frigid ocean. The term "loses itself" is accurate, for in the Mackenzie's many-stemmed journey across the low-lying plain it is almost impossible for the eye to detect where river ends and sea begins. It has traveled far, this father of northern waters, since first we came upon it as the Athabaska, flowing softly on to wed the Peace.

The river of the Indian tribes, the river of the fur trader – and now the river of the mineral hunter who brings the devices of science to his aid to crack open the frontier and partake of its treasure. This, in brief, is the story of this vast Empire of the Mackenzie.

It has long been written in the mythology of the outside world that it was the white fur trader, pushing into the deep north with his canoe- or sledload of bright beads and rum, to trade for priceless skins, who "opened the country." Not so. The last interest the trader possessed was that of civilizing the wilderness, for to civilize its people would be to teach the native the true value of his furs – and why would any honest trader entertain such strange ideas? Moreover, as settlements are founded, the fur-bearing and game animals tend to move on. Rather, then, it has been the way of the fur trader to resist the encroachments of civilization and to look angrily askance upon the opening of every new modern community in the wilderness. Nevertheless, the fur barons were the white trail blazers of the Mackenzie, where they held sway in lonely and profitable majesty until the airplane came to the wilderness.

By then it was the middle of the 1920's and the hare-brained innovator was a certain Punch Dickins, World War I flier turned bushpilot, who was determined to show the Canadian government that the feasible way to deliver mail in the Territories was by airplane, not dog team. In the depth of the northern winter Dickins flew a rachitic and elderly Fairchild down to Fort Resolution, where he cracked a propeller blade in landing on the hummocked ice of the bay to the shores of which that settlement clings. Without means of communication with the outside world, Dickins sat down to ponder his problem. So far as his sponsors and friends "outside" were concerned, he was just another lost aviator. But in the back room of the Indian agent's shack at Resolution, Dickins and his engineer whittled and filed until they had brought the two-bladed prop into balance again. Long weeks of painstaking work went by. But, before the ice went out, they took off and flew south to where railroad tracks span southern Canada. And so a new era was born in the Land of the Mackenzie.

The airplane brought the prospectors. They found gold at Lake Athabaska and north of Great Slave Lake in the Yellowknife country. LaBine stumbled across pitchblende at Great Bear and the first wild rush was on. As the engineers and geologists, hard-hitting, modern-minded men, moved their gear into the country, they brought electric light, steam heat, running water and indoor plumbing to the hinterland back of beyond. The Athabaska, the Slave, the Mackenzie flashed blue in summer, china-white in winter, under the pontoons and skis of their planes as they crisscrossed the country looking for favorable geology from the air to test with drill and dynamite on the ground. Instead of dog teams and slowpoke riverboats, planes winged the mail to Providence and Simpson, to the oil wells at Norman, the radium mines at Great Bear and on down to the Arctic coast. The bush fliers and the mining men had finally pried open the door to the wilderness empire, a door which the fur traders had hoped they could keep locked forever.

Fortunately for North Americans and their allies the fur traders' wish was denied them, for World War II was to prove that the northwestern corner of this continent is the soft underbelly of the Western Hemisphere, wide open to any aggressor who comes across the top of the world through the air or the stratosphere. It was the knowledge of the Far North, its weather, its routes of travel and its terrain, which its unsung air pioneers had acquired, sans help from those charged with the defense of country, continent and hemisphere, that provided the bedrock for the hurry-up-and-the-hell-with-expense defense measures the United States and Canada undertook jointly at the outbreak of war. Even so, all that saved northern British Columbia, Alaska and the Yukon from successful invasion and deep penetration of this continent by the Japanese was our enemies' decision to send their main thrust southward, rather than north over the Great Circle against North America. In the beginning it was the Japs who saved us, not ourselves.

Across the wilderness Canada built the chain of mammoth airports of the Northwest Staging Route, leading down to Fairbanks, Alaska, and out across the Pacific. To join them on the ground and to supply defense forces in his own northwestern territory, Uncle Sam built the Alaska Highway, a military road in every sense of the term, despite its advertised tourist attractions. From far down the Mackenzie, the Canol pipe was driven through the impenetrable mountains, only to be written off as a $400,000,000 blunder and abandoned. But, while all this activity was afoot, the northwestern empire was being equipped with landing strips by the dozen, with weather and radio stations, even a beam route down to Alaska. Most of the Army of the United States has long since gone home, but the amenities remain. The north has been cracked wide open as it could not have been cracked in a quarter century of peace, and through the aperture the planes fly down to Yellowknife, Great Bear and the oil wells, complete with blonde stewardesses and hot meals served aloft to parka-clad voyagers over the tundra. The defense problem nevertheless remains and it is North America's most acute military question mark, for the underbelly is even softer in the dawning age of rocket weapons than it was in the era of the four-engined bomber and troop transport. But that, too, is another story.

Though the crack becomes a fissure and the fissure a crevasse, the frontier remains tough, a country where Nature always fights back. The dog team traveling the frozen Mackenzie gives way to the CAT-train – tractor, sleighs and caboose – and the traders' canoe to the tug and its string of barges. But settlements are still from one to three hundred miles apart. The surface of this vast treasure house has still not been scratched. Even though a century of modern attack upon its wealth may yield millions upon millions in new wealth, the result will be no more than to cut a few gashes in its profile. The last great American frontier will always be the frontier.

Of that frontier, the Mackenzie and the tributary rivers and great lakes of its immense watershed will always be the lifeline. The wilderness empire and its majestic river are one and indivisible.

CHAPTER TWO

Before the White Man Came

SOMEBODY IS FOREVER telling us what happened here and what happened there "before the white man came." Often we do not know what he found when he did come, for not all the early visitors who came to the Americas, usually seeking a new route to the Indies and striking the great western land mass at various parts of its huge eastern seaboard en route, were gentlemen addicted to keeping diaries concerning their contacts with the native population.

Many explorations inland differed little in this respect from those which came from the sea and ended near the coast. But Alexander Mackenzie, whose name this great river takes, was not merely a man addicted to the writing of diaries and the keeping of journals. Indeed, he can be better described as having been prolix in such matters, to the extent that he has left behind an assortment of papers from which it is possible to cull a condensed version of what he found in the way of native humanity and how that humanity occupied itself at the time he voyaged down the Mackenzie to the Arctic Sea in the summer of 1789, and so "discovered" the great river.

Before the white man came to the Mackenzie Valley and the huge basin of its tributary rivers, the country was inhabited by Indians of the Athabaskan group of tribes in the south, and by Eskimos in the fringe along the Arctic coast. These tribes belong to the most widely distributed of all the linguistic divisions in North America and are related to the southern Athabaskans of New Mexico and the Pacific Athabaskans, first met in southern California.

So far as the tribes resident in the area of this great watershed are concerned, they fell into several groups. South, and roughly around the area where the Athabaska River flows across the delta between Lakes Clair and Athabaska, were the Chipewyans, for whom were named the great fort and trading post Mackenzie and his fur-trade partners built at the west end of Lake Athabaska, the base from which Alexander Mackenzie set out in 1789. Somewhat north of there, in the Slave River country, lived the tribe known as the Slaveys. Over west, along the Peace River, were the Beavers. But the principal inhabitants along the marge of the Mackenzie itself, from the point at which it flows out of Great Slave Lake well down toward the Arctic Circle, were the Dogribs, and north of these the Hares. These two groups are our principal concern here, because it is of them that Mackenzie wrote in his journals, and their life which he has portrayed for posterity.

In the main they were peaceable enough. At least they gave Mackenzie and his associates no trouble, but tended rather to give a wide berth to the white man and his party. What may be called the social and political organization of communities was obviously scant. They lived along the riverbank in family groups and groups of families, each of which had its own chief, who represented the clan in its dealings with other native groups and, later, with the white man.

Warlike the Dogribs and Hares may have been among themselves, and when the latter came in contact with the coastal Eskimos bloodshed sometimes ensued, with the coast dwellers, however, as aggressors. But in respect to the white man, the attitude was one of awe and timidity. Mackenzie and other early voyagers into the country report that when their canoes came downriver, approaching groups of Indian lodges on the bank, the native inhabitants made no show of hostility and in many cases did not even stay to inquire, but simply took to the woods. Those who encountered his party on its way north were at pains to avoid it when Mackenzie returned southbound, obviously from fear—fear of being taken into slavery, fear of loss of its young women to Mackenzie's southern braves, or just plain, superstitious fear.

They were migrant and they lived off the country as best they could, a country by no means suited to agriculture, of which they possessed no recognizable knowledge anyway. So they hunted and fished with crude but ingenious weapons: bows, arrows, spears and clubs, made from wood, bone, flint, stones, and in some cases copper and small bits of iron. As their principal sources of food and clothing were moose, caribou, beaver and hare, the tribes were almost constantly in motion, following the migrations of the game animals that sustained them. The picture of native life along the Mackenzie before the white man arrived, in short, is that of small parties of braves, women and children, in movement along the banks of the great water, from Great Slave Lake down to the delta.

Mackenzie's own description of his first major encounter with the Dogribs, shortly after passing out of Great Slave Lake and entering the river proper on his voyage of discovery, is perhaps as good a key as may be found to the life and ways of the native at the time of the first whites' arrival. That was on Sunday, July 5, 1789, shortly after seven in the morning, and the explorer's party had already been on the water five hours, for this was the season of the year in the far north in which night itself is little more than a gentle dimout of day.

Smoke was picked up ahead on the right bank and the Mackenzie party made haste to reach it, but as the canoes approached, braves, women and children were scattering into the woods. Mackenzie's Indian guides from Fort Chipewyan managed to come up with laggards of the scattering party, who apparently were terror-stricken. In fact their terror increased on being addressed by men of their own pigmentation and general habits of dress. Obviously this was due to the presence of whites. The Dogribs had met Chipewyans before, for the latter occasionally had come to trade for fur on the river. But the visitors contrived to create a friendly atmosphere and, as the morning wore along, those who had hidden in the woods began to drift back timidly to the riverbank. By noon their fears had been allayed to the point that the Chipewyans in Mackenzie's party were able to make themselves understood. Fear of the newcomers seems to have finally disappeared when the explorers broke out the rum keg and passed refreshment around, thus inaugurating on the banks of the great river a custom which the white man has carried to the most sequestered corners of the earth in his quest for trade and profit.

Later, the travelers made presents to the Indians of beads, gartering (*sic*), rings, flints and hatchets. After that there was no keeping the Indians out of the visitors' tents. Then, when the gifts had been stowed, the Dogribs put on a show to show their gratitude. As Mackenzie saw it, the tribal dancing was a disorganized business, in which the participants moved around and around in a circle, the men howling in imitation of animals (possibly the effect of the explorers' rum, a cordial which has worked similar results on more than one white man before today, at points much farther south). The visitors were not particularly amused.

"They are a meagre, ugly, ill-made people," says Mackenzie in his journal for that day, "particularly about the legs, which are very clumsy and covered with scabs. Many of them appeared to be in a very unhealthy state, which is owing, I imagine, to their natural filthiness. They are of moderate stature and, as far as could be discovered through the coat of dirt and grease that covers them, are of a fairer complexion than the generality of Indians who are the natives of warmer climates."

Many of the river Indians wore their hair very long. Others cut it short to the head, excepting a long single tress, or queue, falling back over the shoulders. Some of the old braves had long beards. Others had plucked the facial hair out by the roots and were beardless as babies. The braves were tattooed on each cheek, nose to ear, and had pierced the gristle of their noses, to permit the wearing of goosequill, or wooden, rings.

Moose and caribou hides provided clothing along the southern reaches of the river, while reindeer skins clad those whose habitat lay farther north toward the coast. A single jerkin covered the body from neck to knee, often crudely embroidered by the women, or decorated with porcupine quills. The bottom of these garments was fringed and men and women alike wore animal-hide leggings from ankle high up the thigh. "The dress of the women," Mackenzie noted, "is the same as that of the men." The same garment, he noted, did double duty, day and night. Getting up in the morning was simply a matter of moving from the horizontal to the perpendicular position and going about the business of the day.

The Indian lodges of the Mackenzie were extremely simple structures. Poles reached down from a common apex to form a semicircle on the ground and these were covered with bark, not skins, to give crude shelter. These semicircular huts were built in facing pairs with room left between for a common fire. Such dishes as the natives used were crudely carved bowls of wood, bark, or horn. Food was boiled in great gourd-shaped vessels, which seem to have been hollowed roots. Water was heated in these by the process of dropping heated stones into the vessel which, being wooden, obviously could not be put on the fire. Such a container would hold from two to six gallons of water.

For fishing, the river Indians fashioned huge nets, made of willow bark worked into thread by tediously rolling it on the thigh. Some of these nets ran to lengths of 50 or 60 feet and were several feet in depth. Short nets were used in the shallows of the river, longer ones in its deeper reaches, or in back-country lakes. Hooks were made from wood, horn and bone, fishlines from the sinews of animals.

Being primarily hunters, the river Indians were well equipped with crude weapons. Their bows were at least six feet in length, strung with animal sinews or strips of hide. Arrows were two and a half feet long, tipped with barbs of flint, iron, copper – or, when these could not be had, bone or horn. Each was winged with three feathers. A primary weapon was the six-foot spear, usually pointed with about ten inches of bone. The favorite hunting technique seems to have been to catch the prey while in the water and attack it from a canoe by coming alongside and plunging the spear between its shoulders, following through with attacks with crude daggers, which were simply sharply pointed bone, or shaped reindeer horns, with which the wounded animal was literally cut to death as it struggled in the water.

The Dogribs were also trappers of a sort, an item of first interest to the white men who came to the river, for much of what the newcomer had on his mind was fur. Hence the discovery that the natives already knew something about catching fur animals was obviously a start in the right direction. Improve their techniques and equipment, bring along a few beads and a crock of rum and soon you would have another Indian working for you, a fellow who knew nothing of the whimsies of beautiful women in the settled areas of the continent and less than nothing about the habits of effete courts on the other side of an ocean, and so could not comprehend this great yearning for fur. The poor subhuman being (the whites' estimate of the native) had never even heard about the ocean!

So you can picture the firstcomers into the country regarding its crude animal-snaring devices with acute interest. The smaller noose snares they did not even trouble to examine. They were for catching items of food, hares and partridges, with which the country abounded. But there were other and larger traps, the purpose of which was to entangle an animal and hold it securely enough to restrain its departure until such time as the hunter turned up and killed it with his dagger.

The Dogribs probably would never have troubled their

scrofulous heads with trapping for furs but for an odd economic fact. South of the river, back around Great Slave Lake and up in the country of the Chipewyans, lived Indians who occasionally ventured north, bringing with them small pieces of metal which made ideal tips for spears and other weapons. These southern Indians acted as middlemen, which they were called, between white traders and distant tribesmen. Such bits of metal the visitors always seemed anxious to trade for furs and the Dogribs, to whom the idea that a dozen beautifully furred pelts could be equal in value to a precious copper or iron spear-tip was undoubtedly nonsense, were happy to oblige. During the long intervals between visits from southern Indians, the river people trapped and stored such fur as they could, in what spare time they had from the urgent business of sustaining life. They then waited in hope for the day when a canoe would appear from the south, driven downriver by some strange-minded red brother anxious to swap precious metal for worthless animal skins. On these terms, obviously, each was the gainer, for each came by what he most wanted. The Dogribs were soon to see the day when the trades they made were less even, for once the white man came and the river people began to learn his ways and come by the invaders' own acquisitiveness, outlooks would change. But, for the moment, the idea of furs for metal was a deal no Dogrib in his senses could comprehend; and in more than one tepee along the river merry *ughs* were grunted over the childish desires of these effete Indians who came down from the south.

For all the slightly supercilious writings of the early white visitors, these were none the less an ingenious folk. Their problem was that of survival and to meet it they had learned to adapt to use such things as nature provided in a country of sparse vegetation, unproductive soil and rugged climate, a country into which winter comes in October and which it does not leave until deep in June. They knew how to take a piece of brown or gray stone, six to eight inches in length and two inches thick, and wear it down to a cutting edge. The resulting axhead was then fastened to a wooden handle with thongs of animal skins. These crude axes were sufficiently effective to fell the slender trees of the region and to split wood for fires.

Everybody, Mackenzie notes, carried a small pouch containing flinty stones and was "in a constant state of preparation to produce fire." The Dogribs' canoes were virtually models for the modern, factory-produced craft, similarly shaped and watertight, made of strips of bark laid on slender poles of fir and, consequently, feather light. Dogrib paddles provided the basic pattern for the modern propellant, with smoothed blades usually about eight inches across, handles five to six feet in length, neatly rubbed to a hand-grip at the top. For what he had to do to keep himself and his family alive, in short, the river Indian was equipped. If he had not troubled to learn more complicated affairs, thereby bringing himself into the supercilious contempt of the white men who "discovered" him, it could be argued that the things he required to know he knew.

It was after the white man came to the river to teach the

native covetousness and greed that the Dogrib tended to become less the simple human animal, more the craven, and inevitably, the white man's economic slave. To this day survivors of the river tribes who have remained at a distance from the trading posts retain far more of the native personality and way of life than do those who have set up housekeeping in and around the white settlements, to learn the white man's ways and customs, absorb a smattering of his education, religion and mores, his habits "good" and "bad." But that is the story not merely of the Mackenzie country, but of what is called White Supremacy, no matter where its banners have been carried. Yet the story of the red man has its place in such a record as this, for it was the red man's river for centuries before it was the white's.

Not even this brief record would be complete if it failed to refer to what moderns would call the Status of Women among the river tribes. In truth, the relation between the sexes does not appear to have varied greatly from that of more refined civilizations, then and now. True, women were the "property" of their menfolk. All the early explorers into what are now Canada's Northwest Territories recorded this as the basic condition of family life. But women also had a great deal to say about what their men did, and Mackenzie reports in his journal that "though the women are as much in the power of the men as other articles of their property, they are always consulted and possess a very considerable influence in the traffic with Europeans and [in] other important concerns."

Marriages were made at extremely early ages, judged by

the white man's standard. That was obviously because there was no economic problem, other than that of survival, which was common to all. The teen-age brave, in fact, might be said to have been better equipped to meet the vicissitudes of life than the older man in his twenties or thirties. He could run faster, paddle harder, plunge a native dagger deeper between the shoulders of a struggling animal. Thus, once he had learned the simple techniques of hunting, fishing, fire-making and the building of shelter, he was ready to become a family man, and did.

The river Indian was a polygamous fellow, untroubled by divorce laws, religious scruples, or the payment of alimony to a discarded wife. Wife-trading was widely practiced within the migrant clans that moved north and south with the seasons and the animal life of the country. But there is ample evidence to sustain the statement that for his own household, the contents of which might change from time to time, the brave took full responsibility in respect to food, shelter and clothing. In short, though surface customs may vary as between people, places and periods, they were and are identical in fundamentals.

Here and there in the early records of the country the investigator comes upon notes concerning the inhumanity of the tribesman toward the aged, ill and infirm, often left behind to perish when the clan moved on. But the reasons for this callous attitude, again, are explained by the circumstances of life and the country in which it was lived. The river people had no beasts of burden. As nomads, when they moved north or south with the migrations of food animals, whatever they took with them had to be carried on the person or in canoes designed primarily for single or at most double occupancy. They had few ways of preserving food to carry over long periods of movement. Thus nonproducing persons, excepting babes in arms, were discarded.

Concerning the Eskimos on the coast, little is written in the early journals that bears on the Mackenzie country proper, for the early white visitors saw none of them. Although they sought them on the islands offshore, and found many signs of recent occupancy, they could find no Eskimos. It is apparent, however, that the river Indians gave their coastal neighbors as much room as circumstances would permit. Only when absolute need for food drove them deep into the delta in pursuit of game did the northern Indians risk contact with their seagoing neighbors, a tough and warlike people who brooked no intrusion from the south. These were the circumstances as Mackenzie recorded them after his first brief visit to the coast. The Indians, he said, lived "in terror" of the Eskimos, for to be taken prisoner meant death, since slavery was not economically workable among nomads constantly on the move in search of food; and nomads the Eskimos were.

Such, then, were the circumstances of life along the great river "before the white man came." Without exception those who wrote of the native did so with a supercilious air that clearly indicates contempt for the "barbarian" who, down to the moment of white arrival, seemed to have been getting along all right, or at least in a manner that suited him and his nature.

In fairness it should be remembered that perhaps this contemptuous undertone which flows through all early reports on Indian life derives from the fact that the usages of language were not so technical as they are today; that the keepers of records themselves made no pretensions to authorship, but were simply setting down what they saw as it looked to them and in language they used in ordinary talk. To ponder such matters would involve an exercise in semantics, which is not the purpose of this book. What may be recorded in simple terms (and it is the core of the matter) is that the white men who "discovered" the great river and who first ventured into the huge country that surrounds it saw in the native a simple animal, an inferior being to be kept in ignorance of the newcomers' true intentions and purposes, so that the Indian might be exploited for the profit of the white. All of which duly happened.

CHAPTER THREE

Alexander Mackenzie Finds a River

AT NINE O'CLOCK on the morning of Wednesday, June 3, 1789, a group of whites and Indians, including four women, gathered on the muddy strip of beach that fronts Fort Chipewyan, where Lake Athabaska narrows at its western end to send its waters north to the Arctic Ocean.

The party embarked in a small fleet of canoes. The first, and largest, was manned by four French-Canadian voyageurs, two of whom were accompanied by their wives, and a German, John Steinbruck. Amidships sat a young, dour and determined Scot, by name Alexander Mackenzie, out of

Stornoway in the Island of Lewis, twenty-six years old, but still a ten-year veteran of the fur trade, and already a partner in the North West Company.

In a smaller canoe was an Indian who had come by the title of the English Chief, by dint of earlier voyages in the north in areas dominated by the Hudson's Bay Company. His two wives sat with him. A third craft carried two young Indians, followers of the English Chief. The fourth canoe was loaded with trade goods and provisions, paddled by Indians and in charge of one Laurent Le Roux, a clerk in the North West Company's employ. Le Roux was to accompany the party as far as Great Slave Lake, where he would remain to trade for fur.

Mackenzie and his company were embarking on a voyage of discovery. Before this, other white men had explored the country north of Great Slave Lake. Samuel Hearne had been to the mouth of the Coppermine River on Coronation Gulf. But Hearne's and other expeditions had been into the vast area that lies north and east of Great Slave. Now Alexander Mackenzie proposed to swing to the west and see what he could find.

The Indians spoke of a great river and pictured its tortuous wanderings, now between ramparts of rock, again across huge stretches of barren tundra, at times in long placid reaches, at others as a sharp current falling toward the great north sea. It was a country, they said, abounding in the fur beloved by the white man. Alexander Mackenzie had decided to go and see for himself, for fur was to Mackenzie and his partners what iron ore is to a Pittsburgh steel baron today.

But Mackenzie was something more than a younger fur lord. Somewhere in this great northern vastness, he believed, lay an outlet through the continental hinterland, one by which man could travel over water into the Pacific. And on the water ran the only right of way along which man could transport himself and his trade goods to places known or unknown.

This, perhaps, is the primary item to be noted about Alexander Mackenzie; he had something more on his mind than just fur and the profits to be derived from trade with the Indians. Not so his partners, back east in Montreal. They were hard-bitten men of profit, merchants pure and simple, who wanted all that the traffic would stand. They made their profits and are forgotten, save for the naming of an occasional street in Montreal, the reason for which not one-tenth of one per cent of the population of that metropolis now living would know; whereas the name of Mackenzie will live while our speech endures, as the white discoverer of one of the great rivers of America.

Even so, it could logically be argued that, without the men of money at his back, Mackenzie's name would not appear on the map. Without the drive for trade and profit, undoubtedly he would not have appeared in the northwest. For it was the money and backing of the merchant princes of Montreal that sent him there, and, admittedly, it was Mackenzie's own desire for fortune that had brought him to the site of Fort Chipewyan and sent him down north to find a great river. Nevertheless, he had something else on his mind, something that was the driving force behind more than one tough and gallant spirit of his day, the hope of finding the Northwest Passage. That passage he never found. But he did find the Mackenzie River and the great empire it drains down to the Beaufort Sea and the Arctic Ocean. Incidentally, he made a fortune in the process, if that is important.

Mackenzie had arrived in Montreal ten years earlier, at the age of sixteen, with no capital except his own energy and character. It has been recorded that he promptly "enlisted" in the fur trade, and the term was then no misnomer, for to engage in trading into the northwest was like joining the ranks of a rough and brawling army. In Montreal three groups of determined interlopers challenged the authority of the Hudson's Bay Company to the monopoly of the vast region between Lake Superior, the Rockies, and the Arctic Circle.

Originally chartered by Charles II in 1670 as "The Governor and Company of Adventurers of England Trading into Hudson's Bay," what later became known as the Hudson's Bay Company was founded for the purpose of bartering trade goods with Indian trappers for their fur in what is now the central northwest of Canada.

Its first Governor was Prince Rupert, a cousin of the King, who was joined in the venture by seventeen other noblemen and gentlemen. The original charter gave the Company exclusive trading, judicial, legislative and executive rights in all the lands watered by streams flowing into Hudson Bay, and the Company of Adventurers gave the name Rupert's Land to their vast domain.

To the lords of the Bay Company these interlopers from Montreal were mere poachers. But, as they saw themselves, they were brave and public-spirited men challenging a huge unjust monopoly. Unfortunately, they were also challenging each other; in fact, it was not until they realized that divided they could not make good their challenge that they came together as the North West Company. That was in 1787, two years before Mackenzie set out to find his river. Before that the battle for fur could not be said to have been more than a skirmish.

On his arrival at Montreal, Mackenzie had entered the employ of one of these three challenging groups, that headed by John Gregory. For five years he thus remained in Montreal with little more than the status of clerk. Then he was sent to Gregory's post at Detroit, and it was there that he must first have shown high executive qualities, for at twenty-two he was admitted to partnership and was undoubtedly what nowadays would be called the "bright young man" of Gregory's organization.

It was precisely at this time that internecine war among the three Montreal groups was at its height. In 1785, the year in which Mackenzie was admitted to partnership, another partner, Peter Pond, had deserted to one of the enemies, stirring up strife in the Lake Athabaska country in the process. In this feud John Ross, another Gregory partner, was killed. It then became clear that all those engaged in challenging the power of the Hudson's Bay Company must unite in a common front, or destroy one another and leave the entire north to the Company of Adventurers of England, who insisted that they controlled it all anyway. As a result, the three factions united in 1787 as the North West Company. As a member of this company Mackenzie moved to Detroit, then to Grand Portage at the head of Lake Superior, thence westward, until he came to the site of Fort Chipewyan, which his cousin Roderick had selected as the center of the new company's operations in the northwest. There Roderick had built a fort and station on a long and easily defended promontory. It was at Chipewyan, then, that Alexander Mackenzie had gathered personnel and equipment for his first voyage into the northern emptiness, in search of fur . . . and the Northwest Passage.

The party proceeded 21 miles west from Fort Chipewyan, then veered northwest through the Athabaska delta. On the farther side of it camp was pitched for the night. The next two days saw them move on past the confluence with the Peace, down the Slave River and over the series of portages between what are now Fitzgerald and Fort Smith. Here a trucking road now carries thousands of tons of material every season around the only unnavigable water between the end of steel rails and the Arctic Ocean. On the morning of June 9, six days out from Chipewyan, the party entered Great Slave Lake and ended the first leg of its arduous journey. Down to here, Mackenzie had traveled the well-used route to the north. But from here on, he would be on his own.

Ways have not changed greatly since the day 160 years ago when the explorer entered Great Slave Lake from the south. Now, as then, northbound water travelers begin to move with the first sign of breakup. Now, as then, the first ships often reach the mouth of the Slave to find the huge lake still blocked with ice; though nowadays they know it is there, and so creep down to the river mouth to be on their way as far north as they can, and thus save precious time in a navigation season which lasts only three months. In Mackenzie's case, he and his party pitched camp on a site where cabins had been erected by Le Roux and a man named Cuthbert Grant four years earlier. This would be the party's last touch with "civilization," but, by the look of the lake, it was one which might continue for some time. Not until Saturday, June 20, in fact, did the ice clear sufficiently to permit them to be on the move again. Meanwhile, the Indians fished and hunted, in order that the supply of provisions carried in the Le Roux canoe might not be impaired by the delay. Late that day a passage through the ice enabled the impatient Mackenzie to embark. But six miles onward the party ran into heavy ice again, and camp was made on an island offshore.

On Sunday they pushed on again westward and made fifteen miles. Then the ice, driven by a half-gale wind, forced them to take shelter and pitch camp on an island about three miles off the coast. The next week was spent in

battle with the naked elements. Sometimes the party would move on ten miles, attempting to skirt the fringes of the ice pack, until a changing wind drove them to take refuge on the first island they could reach. At other times they would sit and wait throughout the hours of daylight, which in June in these latitudes means twenty hours in the twenty-four, hoping for a break that would enable them to press on. They passed the mouth of the Hay River and paddled toward the western extremity of the lake; paddled past what is now called Big Island and on into the long-necked area, where Great Slave prepares to pour itself into the Mackenzie. Here they fumbled about, seeking an exit, still fighting ice, wind and occasional fog, landing and waiting, launching and pressing on, driven by the determination of Mackenzie to be on his way.

On Monday, the 29th of June, they entered the river proper, twenty-six days out, the last nine of which had been spent in navigating what is still one of the trickiest stretches of water in North America; tricky by virtue of the fact that flat-bottomed ships, constructed for the shallows of the rivers, must risk more than 150 miles of inland sea as treacherous and uncertain as Erie, or any other lake on the continent. They pitched camp late that day, several miles in from the lake. Alexander Mackenzie had discovered his river.

The explorer-trader was now in country and on water where no white man had preceded him. Ahead, the slender ribbon ran on through hundreds of miles of untouched wilderness to reach the polar sea. Filled now with excitement, with the belief that perhaps he had come upon the Northwest Passage after all, Mackenzie drove his boatmen hard. Ice and strong currents often proved hazardous, but the party pressed on. Often it was against the protests of the English Chief and his Indians, who constantly insisted that dangerous rapids were sure to be encountered around the next bend. At day's end, on July 1, they passed the mouth of the Liard, where that great river comes tumbling off the mountains of the west, and where Fort Simpson, one of the largest trading establishments in the country, stands today.

On July 2 the party made 63 miles, passing many deserted Indian lodges on the bank. On the 3rd only 12 miles had been covered when the canoes were driven ashore by a strong head wind. This held them back until well into the afternoon. While they waited, Mackenzie estimated his position and decided that he had come 217 miles west and 44 north, since leaving Great Slave Lake. During the afternoon the canoes passed through currents so strong that, in Mackenzie's own words, "the waters hissed and boiled." Neither the English Chief nor his followers liked their leader's practice of pressing on in such peril. Nor did they, in fact, care much about continuing the journey. They were now well out of familiar country, venturing into the unknown, and the Indian from Chipewyan felt as strange and uneasy as a New Yorker might understandably be were the latter to find himself alone in backwoods country, country that neither he nor any of his companions had ever seen. The

Indians talked of warlike tribes ahead, of threatening cataracts, of strange spirits. Mackenzie alternately cajoled and commanded. But the sense of trouble, born of fear, was in the air. In this atmosphere the party nevertheless pressed on throughout the day of the 4th, and on the morning of the 5th encountered the Dogrib Indians whose ways of life have been discussed in the preceding chapter.

It was at this encampment that the English Chief's fears and dissatisfaction came to the boil. Only when one of the local tribesmen had been impressed into service as guide were Mackenzie's own Indians grudgingly willing to proceed. Late that afternoon the party passed the mouth of the Great Bear River, down which, almost a century and a half later, was to come the raw material for the bomb that shook the world at Hiroshima. Farther on, they camped close to the site of the oilfields at Norman Wells and immediately opposite the point whence more than a century and a half later the Canol pipeline would be driven through the mountains into Alaska.

The sense of urgency was strong in Mackenzie in these days on the river, a growing urgency which drove him onward, deeper into the north with every hour. The canoes made 74 miles on July 6 after breaking camp at three in the morning, and came to rest close to the Sans Sault Rapids. Next morning they were on the water again by four o'clock, and 20 miles farther on passed the mouth of the Bluefish River, the later site of Fort Good Hope. Here they came to the reach where the river flows between high white ramparts, through a gorge at most 300 feet in width.

Beyond the gorge they came upon still another Indian encampment, all the dwellers of which ran off, excepting one extremely old man and a squaw too weak to move away from the campsite. The ancient brave explained that he was too old to care what happened. But when younger members of the party watching from the fringe of the woods saw the strange paleface handing out gifts of beads, knives and awls to the old man, they came slinking back and soon began to make talk and accept Mackenzie's gifts.

By this time the young guide, acquired two days previously at the encampment south of the Bear River, was showing symptoms of alarm. Fearsome tales of the Eskimos, whom he expected to encounter from hour to hour, rumors which had seeped south over the river grapevine, had firmly settled the youth's mind toward return to his own people. That night, while the camp slept, he deserted.

Next morning, Mackenzie persuaded another brave, one from the latest lodges to be visited, to accompany him north. They had now come into the country of the Hares, and the new young brave seemed willing enough to go along. But almost as soon as they were on the water, he too began to speak of turning back, to complain of illness, and to manifest what nowadays would be called lack of morale. This, as usual, communicated itself to Mackenzie's own Indians. The journey was becoming something more than a mere expedition of exploration; it was now a desperate adventure in which the leader found himself constantly forced to plead with his own Indian people, urgently needed for hunting, to press on with him.

On the 9th Mackenzie came to an encampment of still another Indian tribe, described by his Hare guide as "the Quarrellers." This time, as the whites poled their canoes in toward the shore, only the women and children left camp. The braves massed along the foreshore, sure sign, said the guide, that they were ready to fight. At first they spoke angrily while Mackenzie's canoe led the way to the beach. But when the Scot replied, showing no fear and holding gifts in his outstretched hands, they were quickly mollified and soon showed a desire for friendship. The first sight of a paleface had identical effect everywhere, an effect obviously born of superstitious fear.

Here was a new kind of Indian, and far different in personality from those of the southern reaches of the river. They were a rugged, a broad and healthy-looking people. Their garments were more carefully made, more intricately decorated than those of the braves back beyond the Bear. Leggings were worn so high that, had they been joined at the top, the result would have been trousers. Their hide shirts reached lower, were more deeply fringed; and there appeared to be in the people themselves what the white man regarded as a sense of decency in apparel, particularly that of the women, something not apparent among the Indians farther south.

From this group Mackenzie had no difficulty in securing the services of a new guide, who informed the explorer that the sea would not be met until they had slept ten nights. This young man expressed his derision of all southern Indians and his lack of fear of the Eskimos. Yet as the party approached closer to the Eskimo country, he too, like his brothers from the south, wanted no part of the coastal tribes and was forever talking about rejoining his own people.

On July 10 the Mackenzie party passed through what are now called the Lower Ramparts, slightly north of the trading post known as Arctic Red River, whence the run down to the beginning of the vast delta occupied less than half a day. At this junction, where the great river begins to separate into numerous outlets leading to the sea, argument ensued as to which channel should be followed. The now thoroughly nervous guide suggested the easternmost as the one in which Eskimos were least likely to be encountered, while Mackenzie held out for the middle passage, by virtue of its better appearance. Had they borne west, down what is now the main passage, the party must have passed the site of the present community of Aklavik. This is now a pleasant settlement of church institutions, police barracks, hospitals, schools, and the center of government for the huge Western Arctic empire. But they held to the middle channel after much debate. During the course of it Mackenzie's own Indians urged him to turn about and return up the river. Were they to keep going, the English Chief urged, clearly they could not return to Chipewyan that season. But what obviously underlay his words was the constant fear that warlike Eskimos might be encountered around any bend in the now greatly narrowed stream.

Now with persuasiveness and promise of gifts, now

with demands for obedience, the dour Scot kept his little force in motion. As they entered the delta he was forced to give ground by promising his Indians to proceed north for no more than a further seven days. If they did not come to the sea in that time, the party would turn for home. Thus, on July 10, they had put 46 miles behind them; another 54 on the 11th. Meanwhile, the new guide continued to spread fear among the Chipewyans with tales of white bears and fish which spouted water high in air, monsters to be found in the "big lake" ahead.

Soon after the canoes took the water in the early hours of the 12th, the party came to an Eskimo encampment, obviously recently occupied and showing fresh footmarks in the mud. Sledges were neatly stacked against the huts. A kettle, fashioned from hollowed stone and capable of holding two gallons of water, stood close to a recently abandoned firesite. The Indians became more nervous than ever, if possible, convinced that the dreadful Eskimos might be upon them at any moment. Still Mackenzie pushed them on, until they came out from the delta to the marge of what appeared to be a lake. Into this they thrust their canoes to reach an offshore island, where the Indians believed they would be safer from Eskimo marauders. Here they made camp for the night.

Then a strange transformation took place in the mentality of members of Mackenzie's Indian household, for after he and the English Chief had climbed to the high ground above their camp to survey the country round about, regret was expressed to the leader that they could no longer press on. Perhaps fears were allayed, because no Eskimos had turned up. The record does not say. That night the party slept on dry ground slightly above the beach, leaving their canoes and gear at the watermark where they had landed. Scarcely were Mackenzie and his French-Canadian voyageurs asleep when they were called in haste to help move equipment and to save the canoes, as the water washed in higher and higher from the "lake." Mackenzie had come to, and slept beside, the tidal waters of the Arctic Ocean, without realizing that he had arrived. But had he found the Northwest Passage? That was the equivalent of whatever the $64 question may have been on the shores of the Beaufort Sea in 1789.

Mackenzie and his party remained on their sanctuary off the coast for five days, naming it Whale Island, for the large number of sea-dwelling mammals they saw while there, a name which it retains to this day on maps of the vicinity. On his second day the explorer drove a huge stake into the frozen ground, bearing his name and the number in his party, in proof of his discovery. The remaining days were spent in traveling with his French-Canadian boatmen from island to island in search of Eskimos, but none were found, and on the 18th camp was struck and the party made its way back to the mainland to begin the long journey home.

During their first night on the mainland their guide deserted while the party slept. One of the English Chief's Indians explained the next morning that the young brave had feared that Mackenzie intended to take him into slavery, and that he proposed to make his escape while escape was still possible. What amazed Mackenzie was that the young man had left behind a moose-hide shirt, an extra which the leader of the expedition had given him, and had made his way into the wilderness clad only in the single shirt in which he had joined the party.

The journey upriver was more arduous than eventful. On leaving the delta behind on the 21st, towlines had to be constantly employed to haul the canoes through stiff currents. The weather was sultry, interlaced with thunderstorms that cooled the air only briefly. Few Indians were encountered, although when the camp of the Quarrellers was passed, a brother of the still missing guide was urgent in his questions concerning the latter's whereabouts; and difficult to convince that Mackenzie had not done away with him. On the 27th the party passed up the Sans Sault Rapids, where they encountered Indians who told them tales of another great river over to the west, obviously the Yukon. Intrigued, Mackenzie set out in quest of further information, constantly seeking new bands of Indians who could give him more dependable news. But few were found, and it soon became apparent that even those with whom the explorer had conversed on his way downstream were avoiding meeting again with his party on the journey south.

Moreover, the effect of Mackenzie's keenness for news of the western river had a strange effect on his own Indians. They were convinced that if he came by dependable information he would set off on a new hunt for his Northwest Passage, and sought by every device in their power to prevent his coming into personal contact with the river inhabitants. Even the trusted English Chief began to turn odd in his behavior, and for some days Mackenzie was at a loss as to what might be wrong with his principal native aide. Not until they were approaching the mouth of the Liard was the reason for the Chief's qualms revealed. That night, August 13, Mackenzie summoned the English Chief to his lodge, where the Indian confessed his fears and then burst into weeping and lamentation. Mackenzie plied him with "liquid consolation," with an extra libation to carry to his own resting place "to prevent the return of his chagrin." But the Chief was never quite the same man again.

The next day Mackenzie ventured a short way up the Liard to study its banks, but shortly returned to his own river and resumed his journey south. Food was beginning to grow scarce. The hunters were bringing in little game; the river was too low for fishing with nets. The party proceeded south with all possible speed, and on August 22 emerged into the head of the bay leading into Great Slave Lake. Finding the seas too high for their canoes, they went ashore and pitched camp.

The treacherous waters of the lake were crossed during ten days of storms. These time and again almost swamped Mackenzie's own great-canoe, and constantly forced the party to take refuge ashore, not always easily reached. On the 24th one of the smaller canoes broke in half in the seas, though its occupants made shore. On the 27th, when the English Chief, with his wives and braves, left the main party to strike out for the country of the Beavers, where he proposed to winter, his canoe, in turn, was broken by the seas, with the result that the Indians rejoined Mackenzie in a state of exhaustion, with all their supplies and gear lost to the angry lake.

Mackenzie cleared the lake on September 2, after ten days of dangerous travel. En route they had encountered Le Roux, who reported a prosperous trading season. As the party passed up the Slave River the first frosts of autumn were biting down and snow was in the air again. They crossed the portages on September 8 and 9. Early on the morning of the 12th they entered the western end of Lake Athabaska, reaching Fort Chipewyan at three in the afternoon, after one hundred two days on the northern waters which had taken them to the Arctic Sea and back again. Alexander Mackenzie had found the river that remains his to this day.

CHAPTER FOUR

Northwest Passage

THE WHOLE STORY of exploration and discovery throughout western and northern Canada is rooted in the two words that are the title of this chapter. Even before interior exploration began, Cabot, Cartier, Hendrik Hudson came seeking not what is Canada at all, but a water route through or across the top of the continent. Those who followed them by driving into the interior north and west had a similar purpose at least in the back of their minds, though perhaps a profitable trade in fur may often have occupied the forefront. But as the search for the passage turned northwest toward the Pacific, even before Alexander Mackenzie's discovery of the great river that bears his name, as well as the days after his memorable voyages, this fact becomes crystal clear to the historian: not only the river itself and the terrain immediately around it, but the whole great basin that forms the Mackenzie watershed and even the vaster area of what is now Canada's Northwest Territories were all laid open and the blanks on the map of the north filled in by men who sought a water route through or over North America to the Orient.

Some of these men came down from the south, as in the case of Samuel Hearne, the Hudson's Bay Company factor who preceded Mackenzie into the country. But he swung eastward before turning north to the coast, and instead of veering west and finding the Mackenzie, after two unsuccessful expeditions in 1769 and 1770, he finally drove through to the Coppermine River and the Arctic coast at Coronation Gulf in 1771. In doing so, it is true that Hearne moved far from the course of the great river and to waters which, in fact, flow away from it. But the country through which he traveled has in large degree become an economic tributary of the vast empire of the western Territories, and Hearne's discoveries, in turn, put up signposts for others to follow. These led his followers into the Mackenzie watershed and enabled them to play roles of paramount importance in opening the country. Hearne, like all the others, sought the Northwest Passage. Failing to find it, he none the less placed new pinpoints on the map of the continent.

Other explorers came by water, through Hudson Strait, and sailed on west across the polar seas in their search, notably Sir John Franklin and his aides. Franklin was followed by several who sought him in vain, after his last and fatal voyage. These finally found the passage, although they only crossed it afoot on the ice. But here again the search for the through waterway played its great part in the opening of the Mackenzie country, for of Franklin's four expeditions in search of the outlet to the west, the first three were carried out overland and were launched into the northern emptiness from Fort Chipewyan; down the river Mackenzie had discovered a brief time before.

Hence it becomes clear that the name of the river and the search for the Northwest Passage are all but indivisible, and that to follow closely early developments in the million square miles that is the Mackenzie northwest it is necessary to etch in not merely the record of those passage seekers who actually came this way, but of those as well who may be said to have appeared only on the fringes of the Mackenzie country, but left their impress on it none the less.

Of these Samuel Hearne was the first.

As early as the beginning of the eighteenth century, Indians had brought copper to the fur traders' fort at Churchill on Hudson Bay, telling stories of large deposits to be found at the mouth of a great river which flowed to the Arctic Ocean. These repeated stories aroused the curiosity of officials of the Hudson's Bay Company, and several expeditions were sent out in search of the river and its copper, always in the hope that in discovering the river they might discover the Northwest Passage as well. The first of these expeditions, under Captains Knight and Barbour, set out from England in two ships, *Albany* and *Discovery*, which were wrecked on Marble Island, near Chesterfield Inlet, off the northwest shore of Hudson Bay. All hands were lost and their fate was not discovered until a half century later. Other expeditions were sent out along the shore of the bay in the course of the next fifty years, and in 1767 the remains of the Knight expedition were found on Marble Island. One of these expeditions had included a Richard Norton, who claimed to have found the Coppermine River. He was awarded £15 by the company, "for the great hardships he endured while travelling with the Indians," but his claim to discovery is vague.

Moses Norton, said to have been the half-breed son of Richard Norton, was governor of Fort Prince of Wales in the 1760's. He was interested in the Northwest Passage and in the Coppermine story, to such an extent that he visited London and laid plans before the company for an overland expedition to the Coppermine. The idea was approved and he returned to Canada and Fort Prince of Wales.

Samuel Hearne was selected by Norton to make the Coppermine attempt. Hearne was born in 1745, the son of a waterworks official in London. After several years at sea in the Royal Navy, young Hearne joined the Hudson's Bay Company, serving as a mate in one of its sloops in the arctic. Hearne had a reputation for being of an adventurous nature and was known to be diligent and accurate. But apparently he was not a strong or forceful personality.

On November 6, 1769, he set out on his first expedition,

with a party composed of two Englishmen employed by the company, two Crees, and a picked body of Chipewyans under their leader Chawchinahaw. He was well equipped, taking supplies for two years. But Chawchinahaw had little enthusiasm for the trip. After doing what he could to discourage Hearne, he deserted him, taking his Chipewyans with him, and leaving Hearne to make his way back to the fort, a distance of 200 miles, as best he could. Hearne arrived back at Fort Prince of Wales a month after he had left.

Undismayed, he set out on a second attempt on the 23rd of February, 1770. This time he left his former white companions behind, as they had proved to be a burden on the first journey, and took with him three Chipewyans and two Crees. His party moved gradually northwest, as his Indian guides led him about searching for caribou and muskox. Late in July, when they had reached Dubawnt Lake, the Indians informed him that it was too late to continue farther, and advised him to winter with them and push on in the spring. He agreed to this, but shortly afterward broke his quadrant, and being left without proper instruments, decided to return to Fort Prince of Wales. He arrived on November 25, nine months after setting out. His second attempt was thus a failure, and had as its only redeeming feature the discovery of Dubawnt Lake. Yet his second disappointment left Hearne more determined than ever.

On his return from this second journey Hearne had met a Chipewyan named Matonabee, who gave him invaluable advice on how such an expedition should be organized.

Matonabee even volunteered to accompany him, provided he be allowed to manage the affairs of the party, while Hearne acted as observer and surveyor. Matonabee insisted that women were indispensable, since they were needed as beasts of burden and cooks. Moses Norton, who combined a severity toward the morality of others with complete indulgence to himself, objected to having women accompany the expedition. His objections, however, were overruled.

The party set out on December 7, 1770, and moved slowly westward, until spring. Matonabee pointed out that, since they must live off the country, it was best to wait until the herds turned north in the spring and follow them, thus ensuring a plentiful food supply.

They reached Cowey Lake in May, 1771, where they were joined by almost two hundred Chipewyans en route north to attack their hereditary enemies, the Eskimos. By May 30 the party reached Pishew Lake, where they left the women and baggage behind. Hearne and his party were ferried across the Conga-ca-tha-wha-chaga River by Copper Indians who guided them over a difficult but well-worn path across what the Indians called the Stoney Mountains, to the Coppermine. The trail proved a difficult one and the party met a severe snowstorm on July 6. But Hearne had found his river.

On arriving, the Chipewyans sent scouts ahead to find the Eskimos and discovered that they were on the west side of the river about twelve miles down, in large numbers. Early next morning the Indians fell on the Eskimo encampment and caught its inhabitants by surprise, while they were still asleep. The victims ran naked toward the river, since they were cut off from the land, and, as they did not take to the water, all were killed. Hearne had tried to dissuade the Indians from this attack and then to restrain their excess, but without success. When Sir John Franklin visited the scene of this massacre in 1821, he found corroboration of Hearne's story in the form of human skulls bearing marks of violence. He named the place Bloody Falls, which name it carries to this day. Obviously Hearne's Chipewyans from the eastern country did not share the fear of the Eskimo common to the natives Alexander Mackenzie was to encounter to the west a few years later.

Hearne followed the river down to the sea and turned back. He found the copper deposits small and too inaccessible for any commercial purpose.

As to the Northwest Passage, Hearne reported: "The continent of America is much wider than many people imagine, – when I was at my greatest western distance, upward of 500 miles from Fort Prince of Wales, the natives, my guides, well knew that many tribes of Indians lay to the west of us, and they knew no end to the land in that direction; nor have I met with any Indians, either Northern or Southern, that ever had seen the sea to the westward."

Hearne returned by a westward route, crossing Great Slave Lake, the first white man to do so. He moved southward for 90 miles, up the Slave River, then turned eastward, reaching Fort Prince of Wales after a journey of 18 months and 23 days, during most of which he and his party had lived off the game of the country. Another pinpoint had been made on the map of the Far North, nearly eighteen years before Mackenzie was to find the great river, over in the west.

To turn backward briefly, it could be said that the search for the Northwest Passage began more than a thousand years ago, when the subject is believed to have been considered by the Vikings. But the net result of their voyages, as far as modern Western civilization is concerned, was the discovery of Iceland and Greenland; all else was lost. Packed into the centuries that followed are stories of adventure which rank with the most courageous achievements of man, strange legends of semimythological nature and accounts whose stark realism is calmly set forth in the pedestrian language of innumerable journals. The milestones in the search via the northwest for Cathay, the magical land of the Grand Khan and its wealth, were erected only after years of almost fantastic struggle with nature and the elements. In many cases they were represented by the skeletons of Englishmen, Dutchmen, Frenchmen, Italians, Spaniards and Portuguese, found long years later by voyagers to whom the high risk of death could be no barrier. Finally, in 1854, Captain-Robert McClure searching in *Investigator* for the missing Franklin expedition, completed, in a technical sense, the Northwest Passage. But to do so he and his crew were forced to walk across the ice. Actually it was not until 1905 that Roald Amundsen sailed through in his little ship, the *Gjoa*, thus closing a chapter of adventure which had its beginnings in the voyages of Erik the Red and those of Lief Erikson hundreds of years before.

In the first attempts, says Cyriax, writing of Franklin's last expedition, "it was hoped that the Passage would provide a shorter route from England to the East (so much desired for commercial purposes) than any then in use, but during the seventeenth and eighteenth centuries it became apparent – largely owing to the enterprise of British explorers – that ice conditions rendered it useless as a commercial highway and that geographical and other scientific discoveries were almost the only benefits likely to accrue from its navigation." Nevertheless, even in a strictly commercial sense, the efforts of the men who sought the northwestern gateway to the East had not been in vain. Discussing the question of motive in his recommendations to the Admiralty for the fatal Franklin expedition, Sir John Barrow, the "father of Arctic exploration," had the following to say: "The utilitarians were at all times ready enough to ask 'cui bono?' but Elizabeth and her ministers, with their enlightened minds, sought for 'knowledge', the result of which they need not be told was 'power'. Observe what followed; the knowledge gained by the Arctic voyagers was not thrown away. Sir Humphrey Gilbert, by his grant of the Island of Newfoundland, made his voyage thither, in which he nobly perished, but his knowledge did not perish with him, on the contrary, it laid the foundation of the valuable cod fishery, which still exists. Davis, by the discovery of the strait that bears his name, opened the way to the whale fishery, still carried on, and Frobisher pointed out the strait which conducted Hudson to the bay that bears his name, and which gave rise to the establishment of the Hudson's Bay Company, whose concerns are of that extensive nature as to be carried across the whole Continent of America and to the very shores of the Polar Sea.

"Lastly, the discovery of Baffin, which pointed out, among others, the great opening of Lancaster Sound on the western coast of that bay which bears his name, has in our time been found to lead into the Polar Sea, through which the North West Passage from the Atlantic to the Pacific will one day be accomplished, and for the execution of which we are now contending, and which, if left to be performed by some other power, England by her neglect of it, after having opened the East and West doors, would be laughed at by all the world for having hesitated to cross the threshold." England did not neglect her mission, and, as a result, the intrepid Sir John Franklin sailed over the threshold with 128 officers and men, never to return. But the search for the remains of this gallant company, extending over a quarter century, actually completed man's knowledge of land and sea relationships in the frozen Canadian north, which is what interests us here.

From the point of view of the discovery of the Mackenzie River and the Northwest Territories, Franklin's last voyage was significant more as the culmination of previous exploration than as a factor making direct contribution to man's knowledge of the boundaries and possibilities of that area of the arctic. Up to the time of Hearne's visit to the Coppermine the limits of polar exploration had been established as the western coast of Hudson Bay. Inland, north of Lake Athabaska, nothing was known to the white man. Some believed the land extended north to the Pole. Others again insisted on the idea of a water passage leading out of Hudson Bay across North America to the Pacific. Hearne had blasted both concepts. There was no Northwest Passage from Hudson Bay, and there was a Polar Sea at the mouth of the Coppermine. Twenty years later Alexander Mackenzie returned from his descent of the river with evidence of a continuous coast line and a northern sea to corroborate what Hearne had seen. Combine these discoveries with those of Bering and Captain Cook, who penetrated as far to the east as Icy Cape in 1778, and some general concept of the western arctic as it was known in 1800 can be formed. Now there was a lapse of time until 1817 before any action was taken on this new presumption. England's ships were too fully occupied in war to be spared for exploration.

But in 1817 the Napoleonic Wars were over, and England found herself in possession of the greatest navy in the world, unchallenged by anyone and needing employment. In this year, too, reports of a southward shift of the ice pack, clearing many previously blocked arctic channels, were brought home by the whaling fleets.

Our stage can be considered set with mention of one other factor: three acts of the British Parliament to encourage the making of discoveries. The first of these, passed in 1745, provided a bounty of £20,000 for the discovery of a Northwest Passage via Hudson Strait. In 1776 the original conditions were modified by abolition of the necessity to enter through that inlet, and an additional £5,000 was voted for reaching, by sea, the latitude of 89 degrees north. Finally, in 1818, further enticements were offered in the form of rewards for partial results, the major prize being maintained as well. In 1828 these acts were repealed, but on several occasions public funds were voted for outstanding contributions to the solution of the puzzle of the north.

In 1818 begins, then, the last chapter of arctic discovery. That was the year in which the then secretary to the Admiralty, John (later Sir John) Barrow proposed "that one expedition should try to sail between Greenland and Spitsbergen in the direction of Bering Strait and pass as close as possible to the North Pole, and that the other should seek a North-West Passage through Davis Strait."

The first expedition was commanded by Captain David Buchan, with Sir John Franklin, then a lieutenant, as second-in-command. Heaviness of the ice made their effort to reach the Pole impossible. The second expedition, under Commander (later Sir John) Ross, with William Edward Parry as his second-in-command, passed through Davis Strait and is believed to have reached Lancaster Sound. Here Ross reported that he saw a great mountain barrier which prevented further westward penetration. After some minor investigation, the party returned home. The mountain range, which

Ross called the Croker Mountains, was later established to have been an optical illusion, a fact confirmed by Parry, who commanded a second Admiralty expedition to this area in 1819-1820. On this occasion the latter entered Lancaster Sound, found that the Crokers did not exist, and, after exploring Prince Regent Inlet, pushed west as far as Melville Island. Here he wintered, his two ships locked in the ice, and in the spring endeavored to push farther west. Ice frustrated his attempts, however, and he returned to England. Parry had proved the existence of navigable channels from Baffin Bay to Melville Island. He had discovered the extensive archipelago known by his name. He had shown that explorers, adequately equipped and provisioned, could winter in the polar regions in safety and comparative comfort. On his return to England, Parry received a reward of £5,000 for having passed the 110th meridian of west longitude.

A year later Parry was back in the arctic again, this time in an endeavour to find the Northwest Passage by following the continental coast line (1821-1823). He passed through Hudson Strait, explored the east coast of Melville Peninsula and examined Fury and Hecla Strait, but was unable to penetrate farther. He returned to find himself promoted to captain, and was shortly after appointed to command another expedition in search of the passage, and spent the winter of 1824-1825 icebound in the north. Shortly after leaving winter quarters he lost one of his ships, *Fury*, and was forced to take her crew into *Hecla* and return. This was his last effort to find the passage, although he led one subsequent expedition into the arctic in 1827 in *Hecla* and by using sledges from Spitsbergen toward the North Pole. He failed to reach his objective, but he achieved the farthest northern point white men had visited up to that time. He was knighted on his return to England, and remained one of the Admiralty's chief consultants on polar matters during the years of great British activity in the Polar Seas. All of which sets the stage for the coming of John Franklin to the western arctic in search of a passage.

Undaunted by the relative failure of Ross's first voyage, the Admiralty intensified its efforts to clear the mystery of the arctic. In 1819, the year in which Parry's expedition to Melville Island was dispatched, the Royal Navy sent another officer who was to gain world fame as an arctic explorer on a mission down the Coppermine River to chart the coast line eastward from the farthest point of Hearne's travels. Lieutenant John Franklin was born at Spilsby, Lincolnshire, on April 16, 1786, and had entered the Navy in youth. He served at the Battle of Copenhagen in 1801, and later at Trafalgar with Nelson. In 1803 he had taken part in a venture to map the Australian coast, and had been shipwrecked. Then, in 1818, he served under Buchan in the luckless attempt to reach the North Pole via Spitsbergen. Limited though his arctic experience had been, he had nevertheless demonstrated his ability as a leader at a time when Britain was seeking officers for exploration in the arctic.

Franklin was thirty-three years old when he set out

with his party, which included Dr. John Richardson, Robert Hood and George Back, both midshipmen, and a sailor, John Hepburn. Before his departure, Franklin consulted with Sir Alexander Mackenzie, and the advice of this veteran of the Canadian northwest and Hearne's journal of the 1770 expedition were all he had to go on.

Almost two years passed before Franklin was ready to proceed down the Coppermine and begin his work. From Chipewyan, the party had traveled to Great Slave Lake, thence to Winter Lake, where they constructed Fort Enterprise. Here they spent the winter of 1820-1821, and in June started for the Arctic Sea. His party included Canadian voyageurs and two Eskimos. Franklin had hoped to lay in a supply of fish and game during the winter at Enterprise, but results had belied his expectations. Also, "it had been difficult to obtain Canadian volunteers for the expedition, and some who finally consented to go were of poor quality." Furthermore, the expedition was much too large, considering that it must live off the country. In June, 1821, however, the explorers proceeded to the Coppermine River and began their descent of that stream.

They reached the mouth of the river in the middle of July and turned east along the coast of Coronation Gulf. After navigating nearly six hundred miles of coast line, Franklin arrived at Turnagain Point, and, as winter was approaching and supplies were running low, decided to put back to Fort Enterprise. On August 22, he began the return journey, via Arctic Sound and the Hood River, thence across the Barren Lands to the Coppermine and Fort Enterprise.

"In high spirits," says the historian Mirsky, "the party left the valley of the Hood River and faced a march across the Barren Lands. In the face of the oncoming winter it was a stormy wilderness where a sturdy lichen, *tripe de roche*, was all that grew; a sterile plain that afforded neither shelter from the stabbing wind, nor wood from which to build a comforting fire; rarely did the hunters bring in game. Tripe de roche became virtually their only food – an unpalatable, noxious weed that produced in many an enfeebling diarrhoea. For days they plodded on, twenty starving men, abandoning more and more of their equipment, desperately burning the canoes for a little warmth. The country grew more rugged, stony hills cut by stony ravines – obstacles that delayed them, exhausted them. At last, after a month of starving and stumbling, they reached the Coppermine River."

Crossing it, Dr. Richardson, who had attempted to swim, for they had long since burned their canoes, was nearly frozen to death and as a result, though rescued and warmed to recovery, was partially robbed of sensation in one side for a year. Finally a canvas canoe was improvised and the party struggled across the swift-flowing waters and proceeded on their grim march toward Enterprise and food. But starvation had so enfeebled them that a day's march was considered successful if five or six miles were covered.

As the party drew nearer the fort, Back was sent ahead to find the Indians and prepare for its reception. The rest struggled slowly after, the men growing weaker, some barely able to crawl to the fire when halts were called. Finally, circumstances reached such a pass that Hood could go on no farther. He, with Dr. Richardson and Hepburn, pitched a "rest camp" in a clump of willows, Franklin and the remainder planning to push on and send back aid. However, the party had been in motion only a short time when several of the voyageurs declared they could not go on and asked permission to remain with the Richardson detachment. This was granted, and they dragged themselves slowly back to the camp, leaving Franklin and the remnant of the original expedition to fight slowly on to Enterprise. They reached the fort in four days, in badly debilitated condition. Imagine how their joy turned to abject despair when they found the building deserted, no sign of food, not an Indian, only a note from Back saying he had gone on in search of the Indians, and planned, if need be, to try to gain the fort on Great Slave.

There was nothing to do but wait and pray. "Day by day," says Mirsky's record, "their vitality shrank. They gnawed discarded deerskins and bones found among the ashes. It seemed as if their very blood must freeze as they lay before an ineffectual fire while huge gaps in the walls let in 20-below-zero temperature." Franklin saw occasional game, but efforts at shooting it were frustrated by weakness, which made even taking a steady shot impossible. Some time after their arrival they were joined by Richardson and Hepburn, both terribly emaciated, and the pair had a ghastly tale to relate. One of the voyageurs who had left Franklin with three companions, had arrived alone at Richardson's camp, to tell a plausible tale of how the others had died of exhaustion and starvation, though he himself seemed to be well fed and strong. The man's actions became increasingly suspicious. When he went hunting, he carried a hatchet instead of the knife customarily used for cutting up newly killed game, as though "he took it for the purpose of cutting up something he knew to be frozen." This went on until Richardson and Hepburn were certain he had killed his companions and was waxing strong on this fare. Daily the man grew fatter, and daily they became weaker. The ghastly climax came when they returned to camp one day, after gathering tripe de roche, to find that Hood had been murdered. Wasted and defenceless, he had been shot through the head, butchered. It was plain that the voyageur felt that the starving Englishmen were at his mercy, while they knew themselves to be too weak to resist an attack. Their one chance to survive lay in the death of the voyageur. So Richardson shot him. Then he and Hepburn had set out for Fort Enterprise.

There the party waited for Back, slowly sinking into the last stages of starvation. Two of the Canadians died. In the nick of time three Indians, sent by Back, came to their rescue and cared for the white men, whose emaciated bodies and sunken eyes told a story that needed no translating. By the middle of November, they were able to travel slowly to Great Slave. There they found Back. Had Back not chanced on the three Indians on his way out, he must surely have perished himself, for his strength was waning and his only food consisted of a pair of leather trousers, a gun cover, an old shoe, and a handful of tripe de roche!

The explorers slowly regained their strength and by the next summer were able to proceed to Churchill, on Hudson Bay, and thence to England. Thus terminated one of the most tragic ventures into the northland, valuable, no doubt, in the work that had been accomplished; but perhaps even more valuable in the lessons learned by the Englishmen. Never again would inexperienced Europeans set out into the arctic without first having as complete a solution to their many problems as possibility permitted. The dread lessons had been learned well and were to stand Franklin in good stead.

Franklin's second expedition north from Chipewyan was designed to co-operate with Parry, then setting out for the third time to try to find a Northwest Passage by way of Prince Regent Inlet. This time Franklin, Back, Richardson, and E. N. Kendall, a mate in the Royal Navy, were to descend the Mackenzie River. The first two were to trace the coast west from the delta as far as Icy Cape, where Captain Frederick Beechey waited in the *Blossom* to greet Parry should he find his way through to Bering Strait. The rest of Franklin's party would turn east at the mouth of the Mackenzie to map the coast between the Mackenzie and the

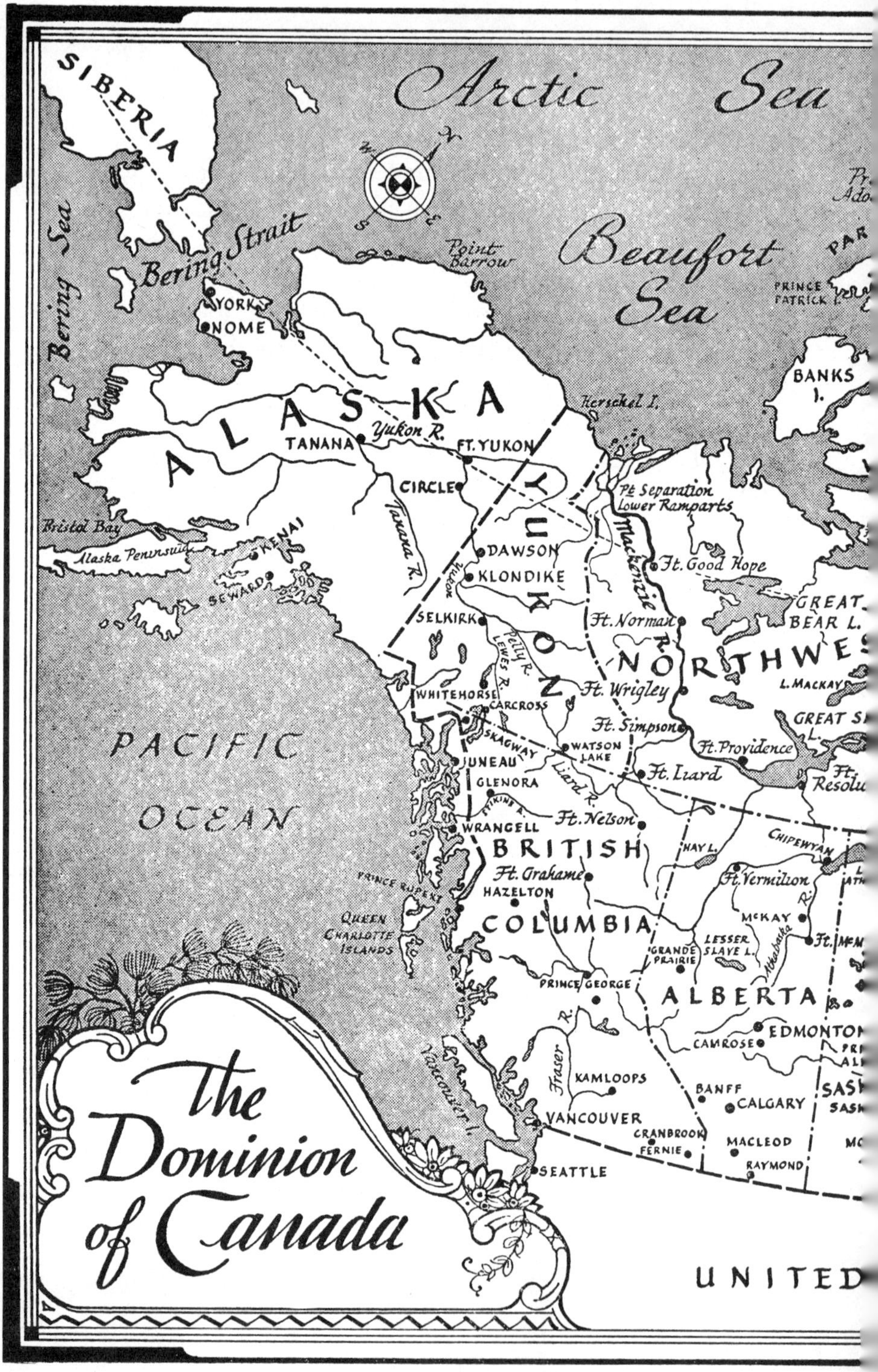
The Dominion of Canada
Arctic Sea
Beaufort Sea
SIBERIA
Bering Sea
Bering Strait
Point Barrow
PRINCE PATRICK I.
BANKS I.
Herschel I.
YORK
NOME
ALASKA
Yukon R.
TANANA
FT. YUKON
CIRCLE
Tanana R.
Bristol Bay
Alaska Peninsula
KENAI
SEWARD
YUKON
DAWSON
KLONDIKE
SELKIRK
Pelly R.
Lewes R.
WHITEHORSE
CARCROSS
SKAGWAY
JUNEAU
GLENORA
WATSON LAKE
Liard R.
Pt. Separation
Lower Ramparts
Mackenzie R.
Ft. Good Hope
GREAT BEAR L.
Ft. Norman
NORTHWES
Ft. Wrigley
L. MACKAY
Ft. Simpson
GREAT S
Ft. Providence
Ft. Liard
Ft. Resolu
PACIFIC OCEAN
Stikine R.
WRANGELL
Ft. Nelson
BRITISH COLUMBIA
HAY L.
CHIPEWYAN
Ft. Grahame
Ft. Vermilion
HAZELTON
PRINCE RUPERT
QUEEN CHARLOTTE ISLANDS
McKAY
Athabaska R.
LESSER SLAVE L.
GRANDE PRAIRIE
PRINCE GEORGE
ALBERTA
EDMONTON
CAMROSE
Fraser R.
KAMLOOPS
Vancouver I.
BANFF
CALGARY
VANCOUVER
CRANBROOK
FERNIE
MACLEOD
RAYMOND
SEATTLE
UNITED

GEORGE ANNAND
ELLESMERE ISLAND
AXEL HEIBERG I.
DEVON I.
N. SOMERSET I.
PRINCE OF WALES I.
Lancaster Sd.
Prince Regent Inlet
Gulf of Boothia
Franklin Str.
BOOTHIA
K. WILLIAM I.
MAGNETIC POLE
COCKBURN LAND
Fury & Hecla Str.
BAFFIN
Baffin Bay
GREENLAND
ARCTIC CIRCLE
Davis Strait
Fox Basin
FOX LAND
RRITORIES
Repulse Bay
Fox Channel
L. GARRY
Chesterfield Inlet
Hudson Str.
LABRADOR
Hudson Bay
CHURCHILL
PT. NELSON
Nelson R.
Hayes R.
YORK FACTORY
QUEBEC
MANITOBA
James Bay
Ft. Albany
LAKE WINNIPEG
MOOSE FACTORY
ONTARIO
St. Lawrence
WINNIPEG
BRANDON
PORT ARTHUR
FORT WILLIAM
Lake Superior
SAULT ST. MARIE
QUEBEC
MONTREAL
OTTAWA
UNITED STATES
ATLANTIC OCEAN
ATES
L. Huron
L. Michigan

Coppermine. Plans were made a full year in advance of this ambitious enterprise, in order that there might be no repetition of tragedy. Before Franklin left England for Chipewyan, he had definite news that all preparatory steps had been meticulously executed. Three special boats had been built, light enough to carry over portages, shallow enough to sail through shoal waters, strong enough to stand up to ice, and stable enough to withstand a rough sea. Franklin intended that the shore of Arctic America should be a continuously known line from Point Turnagain to Cook's farthest, Icy Cape. As to information, again all Franklin knew was contained in Mackenzie's journal of his voyage in 1789. Although this recorded little more than the fact that Mackenzie had followed the river to the sea, Franklin at least knew the location of the coast at two points and to fill this gap seemed a task possible of accomplishment.

The route of the second Franklin expedition followed that of Mackenzie's voyage of discovery, from Chipewyan to Great Slave Lake and down the river past the new Forts Providence, Simpson, and Norman. At the last of these Chief Factor P. W. Dease, of the Hudson's Bay Company, joined them and the entire party proceeded via the Bear River into Great Bear Lake where they constructed Fort Franklin and spent the winter, during which, under Back's supervision, a more-than-adequate supply of provisions was laid in. In June the expedition set off for the sea, leaving Dease behind to supervise activity at the fort in anticipation of their return. This time there should be no calamity like that of Enterprise! They arrived at the mouth of the Mackenzie on July 4, 1826, and divided into the two groups previously agreed upon for exploration east and west along the coast. Striking west toward Icy Cape, Franklin's party encountered frequent heavy ice. "They kept going whenever they could," says the record, "through shoal water, dragging their boats over reefs, rowing in and out of drifting ice, lashed by gales and driving rains, tormented by clouds of mosquitoes, but not until a smothering fog imprisoned them did they stop." After six days in the fog on Return Reef, they gave up and retraced their steps to Fort Franklin. They had been within 160 miles of the advance party sent out by Beechey and had mapped a total of 374 miles of new coast. In September they reached their base, to find that Richardson had arrived two weeks earlier.

Richardson's party had been bothered, too, by shallow waters, gales, rain, fog and mosquitoes, but nevertheless had reached the Coppermine without mishap, mapping in the coast (including Cape Bathurst, Dolphin and Union Strait, Wollaston Land) as they progressed. The river was running far too quickly to attempt to navigate it, so they had cached their boats and superfluous stores and returned to Fort Franklin on foot.

Franklin returned to England in the fall of 1827. In the decade since he had first entered the Arctic Circle, vast areas had been charted by land, in which work he had played a large and important part. The coast line of Arctic America had been traced – save for the few miles separating Point

Barrow and Foggy Island – from Alaska to Point Turnagain; the Coppermine and Mackenzie rivers had been surveyed; Hearne and Mackenzie were at last recognized for the work they had done; and the great blank that had occupied the center of the arctic map had been cut and crossed by thousands of miles of exploration. England recognized the importance of Franklin's achievements by knighting him. Following Mackenzie's trail and extending it, John Franklin had done much to clear the highway to the top of the world, charting much of the basin of the great river in the process.

In the years between 1827 and Franklin's last and fatal expedition, British exploration in the arctic continued. Captain John Ross, in the years 1829-1833, battled with polar elements and, although forced to abandon his ship, *Victory*, achieved several notable feats. On his return, "all England turned out" to welcome him. Ross had redeemed his good name, for he had been subject to severe public criticism after the "Croker Mountains" blunder. His expedition had achieved a record in having passed four winters in the arctic with but three fatalities (two from nonarctic causes); the northern extremities of Continental America had been determined; five hundred miles of adjacent country had been charted, a new land mass discovered (King William's Land); the series of observations that had been kept continuously were the most valuable ever made. The North Magnetic Pole had been located. On his return to Britain, Ross, like Franklin, was knighted and was voted a reward of £5,000 from public funds, despite previous repeal of the acts on bounties.

Other major contributions to man's knowledge of the arctic areas were made during the same period by Commander George Back, of whom we have already read, and by three men of the Hudson's Bay Company, Dease, Simpson and Rae, although the last-named did not begin his explorations until the year Franklin set out on his last voyage. The total outcome of these discoveries was the completion of the map of the coast line from Icy Cape to Simpson's farthest east point, King William's Land.

Before dealing with Franklin's fatal expedition of 1845, it is necessary to examine the activities of men already mentioned in their capacities as aides and helpers of primary figures, several of whom contributed valuable segments to the knowledge of the lands drained by the Mackenzie, Coppermine, and Great Fish (or Back's) rivers, and so contributed in high degree to opening the Mackenzie Empire.

In a day when explorers could not be regarded as missing until they had been out of touch with their own people for a matter of years, Ross had steamed away in *Victory* in 1829. The year 1833 came without word of him, and, as it was deemed impossible for Englishmen to survive four successive winters in the high latitudes, a private expedition was made ready to ascertain his fate. Back accepted command and had with him Dr. Richard King, two carpenters, and a shipwright. The plan was to proceed overland and search the coast. They set out in February of 1833 for Great Slave

Lake, to locate the headwaters of the Great Fish River, which Indians said flowed northeastward to the "frozen sea," in the approximate direction Ross had taken. The Hudson's Bay Company helped Back by furnishing fifteen additional men, supplies, and food. On arriving at the eastern tip of Great Slave Lake, Back left half the men to establish Fort Reliance, the site of which may still be seen, for their wintering, while he pushed on to locate what the Indians call the Great Fish River. Back's preliminary survey ahead was successful and he returned to Fort Reliance to await spring and the breakup.

The winter at Reliance presented an unexpected difficulty in the apparent decision of most of the Indians in the neighborhood to rely on the Back party for sustenance. The Indians soon reduced Back's food supply to the danger point, but the situation was saved by the arrival of Back's old friend (from the Franklin expedition) Akaitcho with food.

Soon after the first signs of spring, and before he was ready to leave, Back received word of Ross's safe return to England. He decided, however, that having done so much it would be an act of stupidity to return from the north with no justification for all the expense incurred in major preparation. For which reason he adhered to his original purpose, and, in July, set out to descend the turbulent Great Fish River, a feat of epic proportions the accomplishment of which would have daunted a lesser man than Back.

From the outset they were beset with difficulties in their battle with nature. The river rolled north with what was described as a "maniacal ferocity of rapids and waterfalls," only here and there marked by a stretch of calm. Often it ran through deep defiles filled with the hollow roar of the water. For days on end, every moment held a crisis. On the 28th of July Back fell in with a tribe of about thirty-five friendly Eskimos who aided in transporting the boat over the last long and steep portage. Back justly remarks in his diary that but for their kindness he could never have reached the sea. Shortly thereafter they came out on a majestic headland, named for Princess Victoria, and saw the Polar Ocean. Of this, Back has the following to say in his journal:

"This may be considered the mouth of the Thlew-ee-choh [Great Fish], which after a violent and tortuous course of 530 geographical miles, running through iron-ribbed country, without a single trace of a tree on the whole line of its banks, expanding into five large lakes, with clear horizon, most embarrassing to the navigator, and broken into falls, cascades, and rapids, to the number of 83 in the whole, pours its water into the Polar Sea about 37 miles more south than the Coppermine River."

The river is now named on the map in honor of the first white man to have made its stormy passage.

Here the record turns to another type of explorer, far different in outlook from the Navy officer solely in search of a passage to the Orient over the roof of the world. These were men of the same breed as Alexander Mackenzie, traders under whose commercial instincts lay the desire to fill in the blanks on the map and to find the passage that was the driving force of all arctic exploration of the period. Peter Warren Dease, Thomas Simpson, and John Rae were all fur men and members of the Hudson's Bay Company. Of them it could be said that they brought to their frontier-busting perhaps a more practical knowledge of the wilderness and its ways than that of others. In any event, they, in turn, added much to the basic knowledge of the region, and that is the important item.

The expeditions of Dease and Simpson occupied three consecutive summers (1837-1839), and on completion not only was the northern coast charted from Icy Cape to the Great Fish River, but the indomitable Simpson had also explored considerable sections of the shores of northern islands, whose presence confronted navigators with narrow channels along most of the eastern section of the coast, where hazards were made doubly complicated by ice floes, fog, and frequent gales.

To expedite their work, Dease had built two 24-foot boats, *Castor* and *Pollux*. Early in the summer of 1837 they set out to fill the gap left by Franklin in his unsuccessful attempt to reach Beechey at Point Barrow. By the beginning of July they reached the Arctic Sea, via the Mackenzie, and during the next two weeks followed the coast Franklin had explored, where, like him, they were bothered by fog, slowed by shoals, blocked by ice. By the end of that month they had arrived at Return Reef, near Foggy Island. Ahead lay the unmapped stretch and the work they had come to do. The going had been hard, slow, and precarious. It was so cold that the fog often turned to sleet, and most of the men suffered from severe colds. Simpson would not give up, however. He determined to proceed to Point Barrow on foot, if possible, and set out with five men, under the necessity of closely following the coast line, because of fog, a fact which increased the distance greatly and necessitated fording innumerable small salt creeks, which kept the party constantly wet to the waist, in water that was extremely cold. The small party rejoined the others without realizing its purpose and repaired to its winter quarters, arriving at the end of September at Fort Confidence in the northeast corner of Dease Arm, one of the huge bays of Great Bear Lake.

During the winter Simpson roved more than a thousand miles seeking the easiest passage to the Coppermine for the next summer's work. The season came late and, with ice still on the lakes in mid-June, Dease and Simpson were forced to drag their boats on sledges overland to the river, so swollen that they were fortunate to reach the sea with their lives. Coronation Gulf, where the Coppermine reaches salt water, was still a solid sheet of white ice and snow. Simpson set out on foot, and in a fatiguing struggle along the coast reached a point where to the east he could see the open water of Victoria Strait, and to the north the loom of Victoria Island. Painstakingly hauling their boats back up the swift Coppermine, they returned to Confidence for another winter.

The next summer luck was better and the warm season came earlier. Down the Coppermine they shot again to find a navigable sea. On this they followed the coast eastward past Point Turnagain, mapping the intricate bays and headlands, and, on August 13, reached the Great Fish River (Back's), thereby accomplishing the major part of their task. Simpson was for pushing on, but a turn in the weather forbade it, so he returned, mapping en route 60 miles of the south coast of King William Island. On the last day of September, in blinding snow, the party regained Fort Confidence after a boat journey of 1,408 geographical miles, one of the longest in the history of arctic navigation.

In the dead of winter Simpson set out for the Red River Settlement, a walk of nearly two thousand miles, seeking company authorization in answer to an earlier request to carry on projected exploration as far as Fury and Hecla Strait. The authority had not arrived. Bitterly angry, not knowing that the desired authorization would be there with the next packet, Simpson turned his back on the north. A few days later he was dead. Whether he committed suicide, as his companions claimed, or whether he was murdered by them, as was suspected, has never been definitely established. In either event, the north lost one of its most brilliant explorers at the tragically early age of thirty-one.

The expedition Simpson had proposed was entrusted five years later, in 1845, to another Hudson's Bay Company official, Dr. John Rae. But whereas Simpson had wanted to work eastward from Castor and Pollux Bay, Rae decided to reverse the direction, hoping to meet the expedition Sir John Franklin was then leading in search of the Northwest Passage. Compared to the Franklin venture, Rae's was a crude organization. Where the British government spared nothing to make its naval expeditions successful, the Hudson's Bay Company expected Rae to do the same work on the smallest possible outlay. He was charged to "determine astronomically all remarkable points, make bearings of all intermediate portions of the coast, chart these daily, attend to all the botany and geology, to zoology in all its departments, to the temperature of the air and water, to the atmosphere, ice, winds, currents, soundings, magnetic dips, and inclinations, Aurora Borealis, refraction of light, ethnographic peculiarities of the Eskimo, and other observations as may suggest themselves to you." In addition, since he had supplies for only four months and the project was expected to consume from fifteen to twenty-seven, he would have to spend much of his time hunting. Ten men were allotted to his party but the all-important part of the expedition was Rae himself; Rae who could outhunt ten men, Rae who could survey on foot more coast line than his two 22-foot boats could sail along. Rae was the first explorer to depend almost completely on the food resources of the arctic, though others had been forced to live off the country when their own supplies ran low.

Rae sailed from Churchill and had reached the head of Repulse Bay at the end of July, 1846. He crossed Rae isthmus to Committee Bay (now Selkirk Bay), where weather and ice made observation and navigation impossible. He and his party wintered at Repulse Bay, resuming exploration at the first sign of spring. He followed the coast northward to

Lord Mayor's Bay (previously reached by Ross) and established the fact that Boothia was a peninsula. After a short rest he then traced the eastern coast of Committee Bay until he was exactly 10 miles south of Fury and Hecla Strait. Then he returned, having "settled the question of a waterway from Fury and Hecla Strait to Point Turnagain, since it was now certain that Boothia Felix was a peninsula and not an island." In all, Rae had etched in the outline of more than six hundred miles of new coast.

Speaking of the work of Rae, Simpson and Dease, P. L. Haworth, in *Trailmakers of the Northwest*, offers the following estimate:

"These expeditions had cost comparatively little; their remarkable success was due mainly to the fact that Simpson and Dease were experienced men who *knew how*. . . . As one follows the long tale of Arctic disasters he realizes more and more that most of them were due to ignorance on the part of those who conducted them. It seems a pity that all leaders were not experienced men such as Simpson and Rae."

It was not inexperience, however, so much as the combination of adverse circumstances that enabled Fate to deal its last hand to Sir John Franklin and the men of HMS *Erebus* and *Terror*.

And now, dark tragedy, yet from tragedy itself final news of the Northwest Passage. . . .

"It is not a little remarkable," wrote George Back in his log, "to reflect on the various ineffectual attempts that have been made by different commanders . . . to fill up the small blank on northern charts, between the bottom, or south, point of Regent's Inlet and Point Turnagain. Parry's and Franklin's achievements are too well known to require eulogium from me; yet the former could not penetrate through Fury and Hecla Strait, and the latter found it impracticable to proceed beyond Point Turnagain. Of Sir John Ross's eventful expedition all have heard. My own, in search of him, is also before the public. Captain Lyon, in trying to reach Repulse Bay (1824) by the *Welcome*, was baffled by a succession of bad weather and heavy gales; and now again, I, acting upon the united experience of most of the distinguished names just mentioned, under circumstances considered favourable, after getting nearly within sight of my port, am stopped by drifting ice, at what is generally considered the very best period for navigating Polar Seas, am frozen fast, in October 1836, at the entrance to Frozen Strait – and now, June 16th, am carried into Hudson Strait, on the very same ice that originally begirt the ship, without having had it once in my power either to advance or retreat. In short, from north, south, east, and west, the attempt has been made, and all equally without effect; and yet with a tolerably open season, the whole affair is within the accomplishment of six months."

Such were the frustrations and problems experienced by the gallant company who sought to force a passage through the frozen northwest!

For a period of seven years, beginning with Back's just-noted unhappy experience at Repulse Bay, the British government had refused to send further expeditions in search of the Northwest Passage. Sir James Ross had been dispatched on three successful voyages to the antarctic in HMS *Erebus* and *Terror* (1839-1843). But not a single ship had cleft the frozen silence of Lancaster Sound on a voyage of discovery. Pressure mounted from a variety of quarters for a renewal of the effort, however: from the Royal Geographic Society; from a public always fascinated by the eyewitness stories of intrepid British explorers; from Beechey, who saw in the newly invented screw propeller the solution to the dangers of seeking the passage; and from Sir John Barrow at the Admiralty. In 1844 the last-mentioned placed in the hands of the Sea Lords a document entitled "Proposal for the attempt to complete the discovery of a North West Passage." Six months later, under the then 60-year-old Sir John Franklin, Barrow's plan materialized when the reconditioned *Erebus* and *Terror*, equipped with engines and capable of making four knots in a calm sea, left England in the most ambitious try for the passage down to that time. One hundred twenty-nine souls were aboard the two vessels when they entered Lancaster Sound. None returned.

The two ships passed through Hudson Strait and reached the Arctic Seas, and, on June 12, 1845, were seen by the crew of a whaling vessel moored to an iceberg and waiting for an opening in the middle ice to cross Lancaster Sound. From that moment Franklin, his ships, and their companions vanished from sight of man. For years no tidings came of their fate.

Not until 1848, three years later, did the Admiralty rouse itself concerning Franklin's fate. For years thereafter, expedition after expedition was sent out to search for the missing men, several by the British government and even one financed by Lady Franklin, who spent practically the whole of her fortune seeking news of her husband.

Even America participated in the search, for the whole civilized world was deeply concerned. Search parties went by ship from east, from west, and overland from the south. Some of these expeditions made important geographic discoveries, and Captain Richard Collinson, sailing from the west, in 1851 reached Gateshead Island, whence he could look across the strait (where, it is now believed, one of Franklin's vessels sank) to King William's Land, where lay the skeletons of some of the men he sought. But of these things Collinson was unaware. Another vessel from the west, *Investigator*, under Captain McClure, reached Barrow Strait, but ran aground and ultimately had to be abandoned. The crew traveled over the ice and joined an eastern searching party, thus actually making the Northwest Passage, though by the ice and not by ship. Nevertheless, the passage had been found and crossed!

It was not until March, 1854, that Dr. John Rae actually obtained authentic news concerning the fate of the Franklin expedition. Rae met a party of Eskimos who told him that every man on the expedition had perished, reinforcing their stories by showing the Hudson's Bay Company explorer objects which had belonged to members of the expedition.

Much later, Captain McClintock, sent out by Lady Franklin, obtained additional information from the natives, found the skeletons of some of the dead and many articles belonging to the expedition, also two short written papers. From all this data the story of what befell the expedition was finally pieced together. It went like this:

Erebus and *Terror* had spent the winter of 1845-1846 at Beechey Island. When the ice broke up next summer they sailed but were trapped by ice in Victoria Strait on September 12, and all efforts to free the ships were in vain. Finally, on April 22, 1848, *Erebus* and *Terror* were abandoned. Up to that time, 24 officers and men had died, including Franklin himself, who passed away on June 11, 1847. Food supply was scanty, and the survivors, 105 in all, set out for Back's River, 250 miles to the south, hoping to subsist on the fish and game they might find there. But it was a hopeless quest. One by one their strength gave out, one by one they perished of disease, cold and hunger. . . . "They fell down and died as they walked," an old Eskimo woman told McClintock eleven years later.

But in one sense, at least, their deaths were not in vain. Search parties to the number of forty-two spent years combing the arctic. They not only finally located the Northwest Passage, but also charted the greater part of what is now known of the western arctic. Later it was realized that Franklin might have completed this task successfully had he only sailed down the east coast of King William's Land. As it was then wrongly believed that this island was attached to the mainland, however, he had attempted to work a passage south toward Coronation Gulf via the western coast of the island, where he ran foul of the ice pack that doomed his venture and ended in death. Yet it is to the work of Franklin that the real credit for final solution of the 1,000-year-old problem of the Northwest Passage must go, even though others who came seeking him found not Franklin but the passage itself.

So closes the great era of exploration in search of the passage, in the search for which gallant men not only established the outline of the western Arctic coast but laid open the vast Empire of the Mackenzie itself. Over in the east, from the tip of Great Slave Lake to the eastern extremity of Coronation Gulf, on the waters of Back's River, they had marked what may fairly be called the eastern boundary of that vast empire. They had found and explored the Coppermine. Great Bear Lake, larger in area than Erie, one day to become the home of the raw material of the atom bomb, had been opened to white human ken. In the west, Mackenzie's own river had become the highway to the top of the world, its waters broken each season by the thrust of the white explorer's paddle. The great day of the fur barons had dawned in the Far Northwest.

CHAPTER FIVE

The Fur Barons

THE HISTORY of the fur trade in the Mackenzie Basin and Canada's Northwest Territories falls logically into three periods, each identified with its own form of transportation. First is the period of exploration and bitter competition, which ended in 1821 with amalgamation of the Hudson's Bay Company and the North West Company, as a partner in which Alexander Mackenzie had traveled down the river to the sea. This, so far as the traders who came from the south are concerned, may be described as the era of the canoe. Second, following the amalgamation, is the period 1821-1869, which could be called that of revived monopoly broadly identified with the York boat in inland transportation — an outsize manhandled craft capable of carrying huge cargoes by comparison with the canoe. The third period runs from 1869 through the 1920's, and marks the decline of the fur trade and its identification, in the Mackenzie empire, with the river steamboat. None of these must be regarded as a sharp division in every respect. Each runs over in at least some degree into the terrain of the other. But in the main the divisions are valid.

Nor is it possible to discuss the fur trade in the region of the Mackenzie and its tributary areas as if it were something apart from the trade as a whole, for events on the Mackenzie were deeply influenced by policies devised in London and Montreal, by the decisions of absentee landlords and by collisions between competitive forces which took place many miles from the silent shores of the great northern waters.

The first date to be fixed in mind is 1670, the year in which what was to become the Hudson's Bay Company was chartered by an English king. That was not the beginning of the fur trade, but of the greatest single organization that has ever participated in it. Long before that time the French, who owned what was known of Canada at the time, were fur traders, gradually reaching westward as their explorers began to open up the interior of the continent.

At first the new company held a distinct margin over its competitors and enemies, the French. It established its forts on the coasts of Hudson and James bays, to which it had access from the sea through Hudson Strait. Its trade goods of British manufacture were superior to those which the French had to offer. The French, in fact, were forced to bring British goods into the country in order to compete. But the competition none the less was conducted with ferocity, even after all military gains of the French were turned back to the company with the signing of the Treaty of Utrecht in 1713.

The French competition ceased after the British conquest of Canada in 1763. After that, Indians in the western areas between Lake Superior and the Rockies could obtain European goods only by long-range trade with the Bay Company at its forts in the eastern arctic, or by paying exorbitant prices (in fur) to Indian middlemen traveling the rivers and lakes of the interior in the company's behalf. Such circumstances invited competition, and it was not long in developing. Independent fur traders and peddlers soon began to cut into the company's trade, by intercepting parties heading north to the Bay forts to trade, outflanking the Indian middlemen. Competition was first renewed in the James Bay and Albany River areas, but soon it cut across the headwaters of the rivers that flowed north into Hudson Bay. As a result the company set out to offset its losses in the south by expansion northwest. Northern Indians were encouraged to act as middlemen, and carry the trade south and west into the Athabaska country and the headwaters of the Mackenzie Basin. Company men were sent out to seek sites for new operations, perhaps the most historically important of these journeys being that of Samuel Hearne, south to the Saskatchewan River from the Hudson Bay coast, to establish

Cumberland House, one of the interior's most important posts and the first to be built by the company; the date, 1774.

The Hudson's Bay Company's attempts to offset the competition of the Canadian traders from the south by expansion northwest was not to succeed, however. Unfortunately for the company, the route of the Canadian traders from Montreal across the inland waterways cut across the headwaters of the important rivers flowing into Hudson Bay. By pushing along the headwaters of the Albany, the Nelson, the Churchill, and ultimately the Mackenzie, the Canadian traders cut across its lines of communication and tapped the sources of the company's trade.

Throughout this period agents were dispatched south by the company to persuade the Indians to come to the Bay forts to trade. One such, Thomas Curry, penetrated as far south as the Saskatchewan in 1771, before Hearne had returned from the Coppermine. Others, Matthew Cording among them, were sent into other areas to persuade the Indians not to trade with the Canadians, but to come to Hudson Bay instead. These attempts availed the company but little, because the direct trade cut into the profits of the Indian middlemen, on the one hand, while the visits of "independents" to tribal areas made it almost impossible to persuade Indians to make long trips across difficult country to the forts. This left the company no alternative but to establish forts in the interior, a beginning in which was made when Hearne established Cumberland House on the Saskatchewan River in 1774.

Still the Canadian traders had the best of it. They intercepted parties of Indians heading to the forts with their furs. Traveling in large parties, pooling their resources, they made free use of rum in their traffic, beginning the debauchery of the Indian tribes. Finally disaster struck from an unexpected quarter when a smallpox epidemic broke out in 1781-1782, in which large numbers of the Athabaskan Indians died, and, according to Hearne, nine tenths of the northern Indians and the company's middlemen perished. The epidemic's effects, of course, were not confined to the Hudson's Bay Company. For two years no furs reached Montreal through agents of the independent traders.

In 1784 a large number of the more important independent traders assembled at Grand Portage at the head of Lake Superior and founded the beginnings of the North West Company, recognizing that competition was harmful to the trade, and that the constantly increasing distances to be covered to obtain furs, the general difficulties of marketing, and the high prices caused by competition made combination desirable. In the arrangement finally agreed upon, a partnership of sixteen members with equal shares was established, in which Frobisher and McTavish, leading Montreal traders, were predominant. But not all the Montrealers entered the New Deal and among the dissidents were Gregory and his partner McLeod, in whose service Alexander Mackenzie was then employed at Detroit.

Thus Mackenzie took service on the side of the smaller and what might be called nonconcurring group. In spite of the general awareness of the dangers implicit in competition, the two Montreal combinations were soon at each other's throats, a feud in which the McTavish group held most of the cards. Its men were more experienced, more numerous, and better financed. Violence soon broke out in the north, with unfortunate effects on trade. The Lake Athabaska region became the scene of bloodshed.

"We now found," Mackenzie records, "that independent of the natural difficulties of the undertaking, we should have to encounter every other which they, who were already in possession of the trade of the country, could throw in our way, and which their circumstances enabled them to do. Nor did they doubt, from their own superior experience, as well as that of their clerks and men, with their local knowledge of the country and its inhabitants, that they should compel us to leave the country to them. The event, however, did not justify their expectations; for, after the severest struggle ever known in that part of the world, and suffering every oppression which a jealous and rival spirit could instigate; after the murder of one of our partners [Ross], the laming of another, and the narrow escape of one of our clerks, who received a bullet through his powder horn in the execution of his duty, they were compelled to allow us a share of the trade. As we had already incurred a loss, this Union was in every respect, a desirable event to us, and was concluded in the month of July, 1787."

This, then, was the business combination that enabled Mackenzie to find his river. The North West Company now had a de facto monopoly on trade out of Montreal, and was free to expand and attack the Hudson's Bay Company at will.

Rapid expansion followed the amalgamation. Forts were built on Great Slave Lake and, in the years immediately following Mackenzie's voyage to the arctic, on the river itself. In 1790, the year after discovery, the first such establishment was built at the entrance to the great waterway and three years later the Nor' Westers moved into Lac La Martre which, though 75 miles north of the river, is nevertheless in its basin and connected directly with Great Slave Lake. In 1796 a fort was erected on the river, 80 miles west of Great Slave. In 1799 the traders moved into Great Bear Lake. In 1804 Fort Simpson was erected at the junction of the Mackenzie and the Liard. Fort Good Hope, beyond the Arctic Circle and adjacent to the Ramparts, was put in operation the next year and Fort Liard, on the tributary river of that name, in the same season. Mackenzie's discovery of a river was beginning to pay off in fur.

At first the Hudson's Bay Company attempted to fight the Johnnie-come-latelies of the North West Company by invading Mackenzie's newly opened territory. A post was built on the Slave River, 25 miles south of the lake, in 1803, and on Great Slave Lake the following year. The river itself was not invaded, however, and in 1806 the older company, finding that it could not compete with traders working into the country from the south pulled out, leaving the empire of the Mackenzie to the newcomers until the second decade of the new century. For the moment Mackenzie's partners had the big river to themselves.

Following the long-standing technique of the older company, the Nor' Westers used Indians as middlemen to open up the trade, selecting post-broken natives already well versed in the acquisitive ways of the whites, and happy to exploit their own people. These southern Indians pushed into the country between and behind the river forts and soon had successfully introduced their simple-minded brethren to the use of firearms, which could be traded for large quantities of valuable fur, as also could mechanical trapping devices and, above all, rum. As the Indian acquired European firearms (which quickly became a necessity to all, for if one group was armed, so its neighbors must be, or perish) and tools, he soon lost his ability with the bow, or to use snares to catch fur with which to trade with the whites. The traders' purpose, not merely here but everywhere, and in the Hudson's Bay Company as with the North West Company, was to make the Indian trapper completely dependent on the trader, thereby assuring to the latter a continuing flow of fur.

The white fur men were divided into three classes: partners, clerks, canoemen. A clerk might rise to the rank of partner, but the boatmen, mostly French-Canadian voyageurs, could not hope to attain rank above that of steersman or guide. In large degree the voyageurs were recruited – shanghaied might perhaps be a more accurate term – while drunk and transported too far into the country before they could recover to be able to escape and make their way back to civilization. It scarcely seems necessary to point out that this was a trade conducted by hard-bitten men who wrote their own laws and made up their own rules as they went along.

The daily rations for each white boatman consisted of a quart of shelled Indian corn and one ounce of grease. There was no halt at noon for dinner. Each crew member's baggage was limited to 90 pounds, and each canoe carried, over and above crew baggage, 600 pounds of biscuit, 200 pounds of salted meat, 3 bushels of beans, tarpaulins to protect goods against rain, a sail, one hawser, one ax, one cook pot, one sponge to sop up casual water, resin, hemp, and birch bark to mend the canoe in case of accident. The craft were often 30 feet long and were always loaded to the gunwales. Over the portages canoemen frequently carried several hundred pounds at a time, while the lords of the fur empire crossed on foot, but empty-handed, excepting perhaps for small personal effects.

After Alexander Mackenzie had completed his voyage of discovery down his river to the Arctic Ocean, he returned to England, and then came back to Canada for a second voyage, this time toward the Pacific. For a considerable time he had been at odds with the senior partner in the North West Company, Simon McTavish of Montreal, and soon became the rallying point of opposition to his rule. McTavish had become the wealthiest man in Montreal, principally through the fur trade, and occupied the dominant position in the company. Both his power and his hectoring manner were generally resented, bringing him such behind-the-hand nicknames as "The Marquis" and "Le Premier."

At the traders' annual meeting at Grand Portage in 1795, Mackenzie bespoke his dissatisfaction with the conduct of the company and several partner-traders actually resigned to join Forsyth, Richardson and Company, a new concern formed to fight Le Premier. Though Alexander Mackenzie did not then leave the McTavish partnership, his sympathies were plainly with the discontented group.

The new combination, usually known as the XY Company, as a result of marking its goods XY, in contrast to the NW of the North West Company, made up in vigor what it lacked in financial resources, but was matched by equal fighting spirit on the part of McTavish. The split came at a time when the Hudson's Bay Company was building forts close to those of the Nor' Westers, apparently determined to re-enter active competition with men it regarded as interlopers. Thus McTavish found his trade threatened on two fronts: by the XY Company, on the one hand, by the reviving activity of the Hudson's Bay Company, on the other. The original plan of the North West Company had been to build forts inland in the north and starve out the Bay Company by cutting it off from the interior, thus making a fiction of the Hudson's Bay monopoly and a reality of its own. McTavish viewed competition as harmful to his interests and was determined to suppress it by fair means or foul. Before the law he had no rights not enjoyed by any other subject of the crown. Therefore the Nor' Westers could only enforce their ascendancy, driving toward monopoly, by force. This they used freely against all active and potential rivals.

An incident which occurred at the head of Lake Superior in 1801 offers a case-example of the times in the fur trade. A certain Dominique Rousseau, of Montreal, sent several canoes loaded with trade goods under the command of his clerk, Hervieux, to Lake Superior, expecting to sell his cargo to Indian middlemen at Grand Portage. Arrived at the lakehead, Hervieux pitched his tent adjacent to the North West Company's fort but on public ground. A few hours later three officers of the company appeared at his camp and one of them, Duncan McGillivray, began by telling him to "get out or they would make him go." Hervieux replied that he had a right to be there, but finally yielded to their threats. As he prepared to move and was nearly packed, the traders returned to his tent and cut his goods to pieces with their daggers, saying, "There's for you! You wanted to see our titles. You have them now. If you try to come inland we will cut your throat!" Such were the ways of the fur trade in the heyday of Simon McTavish.

Rousseau took his case to court but received only slight compensation. Later he made another attempt to enter the trade and, in 1806, sent a young man named Delorme west. Delorme tried to avoid being seen by the Nor' Westers, but was followed by members of the company, who felled trees across the trail before and behind him. Being unable to move his goods, Delorme was forced to sell to the company at cost price and counted himself lucky to get out of the country alive.

Meanwhile Alexander Mackenzie had gone to England to see to publication of his journals, an event which gave him wide renown, brought him knighthood and an invitation to be the traveling companion in Canada and the United States of Edward, Duke of Kent, father of Queen Victoria. Mackenzie's book was so great a success that translations were made into several languages. Napoleon caused a special translation to be made for Marshal Bernadotte, later King of Sweden, who had been assigned the task of preparing a projected invasion of Canada via Louisiana. Mackenzie's voyage down a river leading to nowhere had given him world-wide reputation.

As Sir Alexander Mackenzie, the discoverer returned to Canada to lend his great prestige to the struggling XY Company. Competition became more violent as McTavish went to ferocious extremes to eliminate the smaller company and fought his former partner bitterly.

Competition was unquestionably bad for profits in the trade. The bargaining power of the Indians went up and it became necessary to give them gifts. The Nor' Westers gave them rum, plenty of it. Competition meant needless duplication, increased cost of labor, high prices in bargaining with the Indians, and waste of fur resources. Bad for profits and trade, the bitter competition of NW-XY days was not without its compensations for the Indians and company employees. But the dour personality of McTavish drove him to fight to destroy his rivals, never to settle.

Mackenzie's return to Canada seemed to inspire McTavish with new vigor. Here was a foeman worthy of his steel. He extended the agencies of the North West Company to the south Saskatchewan, over the border to the Missouri, even rented "the posts of the king," as the trading stations on the lower St. Lawrence were called. Since he held the high cards of ample capital, McTavish might well have succeeded, but he died in 1804, thereby removing the principal barrier to amalgamation, which soon followed. Now the full weight of the Montreal merchants could be directed again toward the Hudson's Bay Company.

Mackenzie wanted to unite the Bay Company with the Nor' Westers from the beginning, for he regarded the trade as a natural monopoly and believed that unbridled competition spelled ultimate disaster for all. He believed that if the North West and the Bay Company could come to terms they might organize the trade continentally, and with it operate the fur traffic of all British territories in North America. At this time the stock of the Hudson's Bay Company had fallen to a low price and Mackenzie endeavored to persuade his partners to pick it up, but could not get sufficient support from them to do so, whereas his private means were insufficient to enable him to proceed alone, although he made a small purchase of shares. Thus, since union was impossible, competition with all its attendant evils again became inevitable.

In this new battle for control of the north, the North West Company had superior organization, larger capital, more numerous and better quality personnel, several times the energy of the conservative Bay Company. The latter employed principally Orkneymen, who lacked knowledge of the country. They were handicapped, in turn, by a rigid organization which tied local managers to decisions made in London. Its wages were less, and although, under pressure of competition, a high degree of loyalty developed, many of its servants lacked enthusiasm and frequently deserted to the Nor' Westers for higher pay. Nevertheless, in spite of these handicaps, the Bay Company held the tremendous advantages of a legal monopoly under its charter, as well as enjoying the cheapest line of communications with its market over the sea. Thus in spite of the damage done to its trade by the violent competition of the Canadian traders, the Hudson's Bay Company was still able to maintain a steady dividend of from 5 to 7 per cent. Slow to make use of its natural advantages, by the beginning of the nineteenth century the Old Company began to show life again.

Because the NW-XY imbroglio came precisely when the northwest country was being opened to the white man's trade, the battle, which had its headquarters in Montreal offices and its field base at the head of Lake Superior, was bound to make its impact on Lake Athabaska, on the Slave, on Great Slave Lake, and on the big river itself. This impact was reflected primarily in higher wages demanded by and paid to canoemen and employees at the fur forts, in extortion

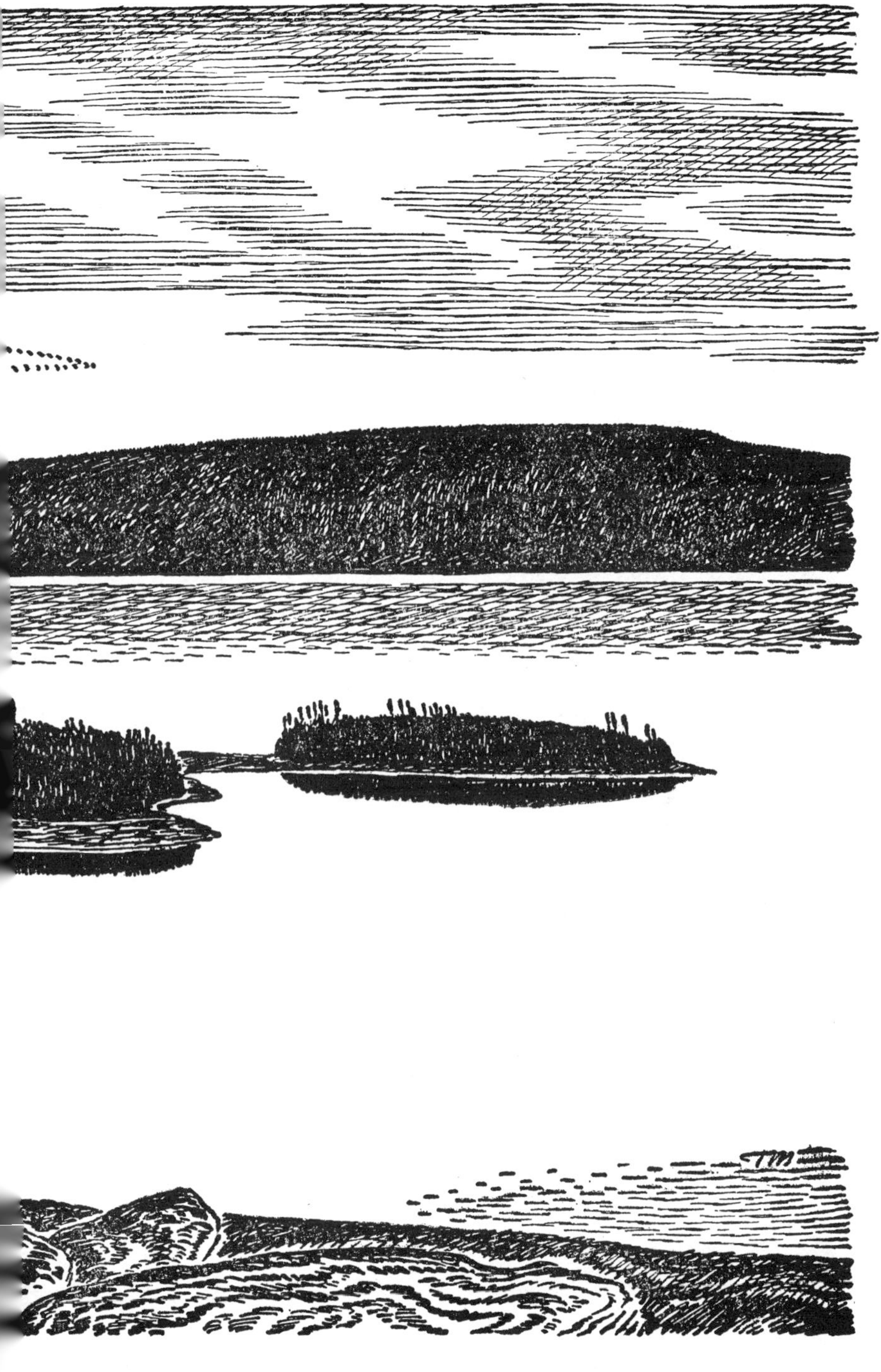

of the warring traders by their Indian middlemen, and south of the river entry, in the prices secured for his furs by the more sophisticated type of native trapper. Second, the influence thrown to the new trading group by Mackenzie, by this time a man of great personal prestige in the northwest, was bound to have the effect of developing a sympathetic outlook on the part of many white men in the country proper, with particular emphasis on the Chipewyan-Lake Athabaska area, headquarters and jumping-off point for the traffic into the north. Third, and finally, it was reflected in the physical aspects of a brawl, in which the principals made their own rules to fit circumstances as they arose, and so could be said to sponsor the reign of lawlessness in the bush. Physical encounters between embattled groups were by no means exceptional in the vicinity of Fort Chipewyan, while on the rivers and lakes of the country white men traveled with their lives in their hands, not under peril from the Indian population but from each other. The point to be made is that the great river was opened to the white man's trade in an atmosphere of what was tantamount to war between lawless men.

Here the page of the Mackenzie story turns to another event of faraway origin, with its beginnings in the highlands of Scotland, which was finally to revolutionize the fur trade in Canada and bring to an end the repeated collisions between warring groups within the trade. When the incident, known as the Selkirk Affair, was closed, the remaining competitors, North West and Hudson's Bay, would join forces through legal amalgamation and bring to a close the era of the fur wars. It was a battle which virtually left both principals for dead, perhaps the only condition in which they could have been brought to treat with each other and finally settle their differences in peace.

Thomas Douglas, fifth Earl of Selkirk, was regarded by his friends as a nobleman of liberal and humanitarian views, by his enemies as a determined and bitter crank. Whichever he may have been, the fact remains that when the large-scale landlords of the Highlands set about evicting tenant farmers in order to turn their land into sheep runs, Selkirk stepped in on the farmers' side: his friends said to help them, his enemies insisting that he sought to exploit the unfortunate. Be that as it may, Selkirk proposed a solution to these unhappy displaced people. The solution was emigration to the colonies.

After reading Mackenzie's accounts of his travels and his descriptions of the Canadian west, Selkirk decided that the most favorable area for the settlement of his Scottish farmers would be the Red River Valley in Rupert's Land (roughly the site and vicinity of the city of Winnipeg). The region, though technically the property of the Hudson's Bay Company under its royal charter, actually was held by the Nor' Westers, who wanted no part of the colonization project and were determined to resist.

Meanwhile Selkirk faced rebuff after rebuff. In London, the Colonial Office refused him permission to settle colonists inland in Canada, with the result that he led his first migration into Prince Edward Island in the Gulf of St. Lawrence, a settlement which, it may be noted, was a tremendous success. Thence he set out on an extensive tour of Canada and the United States, returning to Britain deeply impressed by the fact that many British subjects were settling in the Republic, for no better reason than that they were apparently more welcome there than on the soil of their own adjacent colony, Canada. But still the Colonial Office turned a deaf ear to the Scottish milord, with the result that in 1810 Selkirk decided to change his mode of attack and to bore from within, by purchasing enough shares in the Hudson's Bay Company to control a shareholders' meeting, at which he would propose the Red River scheme.

By May, 1811, Selkirk and his friends had picked up Hudson's Bay shares to a value of £35,000 out of a total capitalization of £105,000. At the ensuing meeting only £45,000 were represented, of which £30,000 was Selkirk money. The result was passage of an enabling motion to permit the Red River settlement to be opened. Opposition came from Alexander Mackenzie and a small group who, it will be remembered, had set about the business of gaining control of the Bay Company several years earlier, but had abandoned the project for lack of sufficient funds of their own, or outside support. Now they came to the meeting to oppose colonization of the fur country, anathema to any trader in his senses. To Mackenzie the very thought of the shareholders of a fur-trading concern voting to bring farmers into its area was appalling. This man Selkirk must be mad.

But, mad or sane, the earl had his way and the great Bay Company formally approved the policy of putting settlers on its land.

Now began, first, open warfare in the Red River country and, second, as bizarre a period in the history of litigation in Canada and Britain as the courts of those two countries have ever known.

The first Red River contingent came by sea in 1811 and landed at York Factory, on Hudson Bay, where late arrival forced them to winter. They had not been long there before the colonists discovered that, although their expedition was officially sponsored by the Bay Company, that company's own factors and officials in Canada were bitterly opposed to their presence. Many of the colonists, idling through the long winter, took the definitely dim view of the snow-girt country in which they were so obviously unwelcome and, had the sea been open and a ship available, no doubt would have packed up and gone home. But their leader, Captain MacDonnel, who had been in Canada previously, was a born disciplinarian, a quality which enabled him to pull his party of seventy people through the winter and, in June, 1812, to lead them out from York Factory towards the site of their new homes where, by autumn, they had established themselves adjacent to the North West Company's Fort Gibraltar.

The settlers' situation was ironical in the extreme. They had arrived in Canada under sponsorship of the Hudson's Bay Company which, according to British charter, owned outright the land to which they had come, yet officers of that company on the spot wanted no part of them or their plans. Moreover, the Nor' Westers were in possession of what was legally Hudson's Bay ground, from which the Bay Company had made no effort to eject them, for the excellent reason that it lacked the physical power to do so. So far as the Bay Company was concerned, it tolerated the Nor' Westers but did not officially recognize their presence. So far as the North West men were concerned, they looked with scorn on the pretensions of the Bay people, in which viewpoint they had the support of the Canadian people and, more important, of the courts in Montreal. Each fur-trade group occupied a legal position which made it right in one place, wrong in the next, the Bay sole possessors of the country in the eyes of British law, the Nor' Westers always favored by the Canadian courts. The immigrants from Scotland were the unfortunates trapped between two warring elements, each making the most of its circumstances.

More colonists arrived. Others left, successfully bribed by the Nor' Westers to give up the project. Those who refused to be bought out were openly threatened with physical violence by the half-breed toughs around Gibraltar. The game ran this gamut until 1814, when the Nor' Westers pounced, arresting MacDonnel and hauling him off to face the courts in Montreal, leaving the Red River Settlement leaderless.

If the Nor' Westers had believed that this move would

disrupt the colony, they counted without Selkirk who, in 1815, in his influential role within the Bay Company secured the appointment of a governor for his settlement, Semple, who came into the country accompanied by a hundred Highlanders spoiling for a fight. To the Red River community were added, as settlers, a number of retired Bay Company traders and their wives and families, many of them part-Indian. Into this unique settlement Governor Semple next brought John McLeod, a Bay Company man who had been posted at the company's nearest fort, on Lake Winnipeg. McLeod set about organizing the Red River people for defense in the nick of time, as events were to prove, for the Nor' Westers, tiring of delays and legal stratagems, had determined to expel the settlers from the fur country by armed force. The first attack was driven off, but the furmen put all undefended houses in the settlement to the torch. Homeless again, the settlers retired to Norway House on the York River.

Now Governor Semple took a hand, turning up on the scene with his Highlanders, leading the homeless back to their clearing, where they all fell to and rebuilt the destroyed settlement, which they now named Fort Douglas, in honor of their sponsor, Selkirk. Now the Bay Company, hitherto none too sympathetic to the settlers' cause, so far as its fieldmen were concerned, entered the fray officially to join the issue with the Nor' Westers. The Bay "army" fell on Fort Gibraltar, capturing and dismantling it. Cameron, the North

West Company factor, was arrested for occupying property belonging to the Bay Company under its charter and was sent out for trial, not in the Canadian courts but to England.

The Nor' Westers fought back. A force was gathered under Cuthbert Grant, son of the Grant who had helped build the first post on Great Slave Lake in the 1780's, and marched on Fort Douglas, bent on battle. As Grant descended on the settlement, Governor Semple moved out to meet him with a hastily gathered force of colonists. A parley was arranged but, as the opposing leaders met, a shot was fired from the ranks of the Nor' Westers, whether by accident or intent has never been revealed. The parley ended before it began, as the enraged Red River men sprang into battle, in which Semple was killed, Fort Douglas was forced to surrender, and the Nor' Westers took over, once again rendering the Selkirk people homeless. The engagement lives in the records of the trade as the Battle of Seven Oaks, bloodiest encounter of all the fur wars.

Now Selkirk turned to the Canadian courts for redress, obviously a mistake since, as the law in Canada then saw them, the Nor' Westers could do no wrong, regardless of circumstances. The principal partners of the North West Company were among the most influential men in Montreal. To Montrealers Selkirk was an outlander and interloper seeking to deprive honest Canadians of their rightful property, a viewpoint strongly fostered by fur-trade leaders in the mercantile town. The courts consistently threw out Selkirk's complaints and refused him redress.

In anger at his treatment at the seat of justice, Selkirk turned to direct action. First he secured his own appointment as chief justice of the Indian Territories. Then, since the crown in Britain refused to send forces to the aid of a private company, he recruited a number of Swiss mercenaries, whom he shipped to the Red River, ostensibly as colonists. This, obviously, gave the Nor' Westers a new talking point, to Selkirk's disadvantage. Who was this Scottish interloper who could recruit private armies to fight Canadians on Canadian soil? Yet, strangely, the Nor' Westers themselves had recruited mercenaries of the selfsame Swiss regiment for similar purposes in the course of the fur war, an item not mentioned in verbal attacks launched on Selkirk from Montreal.

The implacable Scot set out for the west, leading his mercenaries. At Sault Ste. Marie he heard his first news of the Battle of Seven Oaks and that Fort Douglas had been razed. In wrath Selkirk pushed on to Fort William on Lake Superior, where in his capacity as a justice he caused the arrest of three partners in the North West Company stationed there, McGillivray, McKenzie and Fraser, charging them with conspiracy. It was March, 1817, when he reached Red River, where the Swiss caught the small garrison by surprise and quickly retook the fort. The next step was to bring the colonists back, which Selkirk proceeded to do at once. That summer Fort Douglas again began to resume the appearance of a quiet agrarian community.

Now the crown itself, hearing of the continuing disorders in the west, decided to take a hand and quell the internecine feudings of the fur traders. From the east Governor Sir John Sherbrooke dispatched a company of militiamen under Colonel Coltman, for the purpose of restoring property to its lawful owners. (The record does not state whether, in Sherbrooke's opinion, this meant the Nor' Westers or the Bay Company, whose tenants the colonists were, in the eyes of English law.) But Coltman traveled west, obviously convinced that in Selkirk he would meet a tartar of a man, a strong-willed and unreasonable nobleman who would probably have to be quelled in person, before order could be brought out of the chaos of Red River. On the ground, however, Coltman changed his mind. His report to Governor Sherbrooke was strongly in Selkirk's favor, even to the point of recommending that any charges laid against the earl in the courts by his opponents should be dropped.

With order restored in Red River, the battlefield again shifted to the courts of Upper and Lower Canada. Here Selkirk became the victim of what many historians and record-keepers of the day clearly regard as a legal conspiracy, some even to the point of accusing the partners in the North West Company of controlling the courts. Fine after fine was levied on the noble Scot, on all manner of trumpery charges. In reply, case after case was leveled against his enemies by Selkirk, only to be thrown out of court. Meanwhile his own personal fortune was rapidly dwindling under the series of extortionate fines.

Selkirk himself put his thoughts down on paper in 1818,

at the height of the legal battle, when he wrote: "To contend alone and unsupported, not only against a powerful association of individuals, but also against all those whose official duty it should have been to arrest them in the prosecution of their crimes, was at the best an arduous task and, however confident one might be of the intrinsic strength of his cause, it was impossible to bear him up against the swollen tide of corruption which threatened to overwhelm him." This was the kind of mix-up in which no holds are barred.

Broken in health by the strenuous life he had been called upon to lead on the fringes of civilization and by the strain of the running battle that had endured for more than a decade, Selkirk returned to Europe to attend to his affairs, and died in France on April 8, 1820. In fighting the Montreal fur traders he had come to grips with the whole force of government and justice in Canada, aligned alongside influential men of its own community. His enemies (as Selkirk saw them) ranged from judges on the bench in Montreal to leaders of the political Family Compact in what is now Ontario, and was then Upper Canada, led by the redoubtable Dr. Strachan. Yet if Selkirk did not win, neither did the Nor' Westers, for the years of embroilment left both companies financially exhausted at the close of a period in which each had paid far more attention to its struggle with the other than to its reason for existence, the trade in furs with the west and north. Financial exhaustion brought the contestants together in 1821, on March 26, in which year an agreement of amalgamation was signed. In the division of shares, the

Nor' Westers acquired a greater number than did the original Bay Company members. But the former's name disappeared, while that of the older trading concern was continued, in order to retain the royal charter. Now the focus of control was to move away from Canada again and return to its original site in London. Inadvertently Selkirk had brought about an amalgamation which, left to its own devices, neither principal would have sought with the other.

So ends the era of the fur-trade wars. Of the developments of the period in the actual area with which this book is concerned little is known. Forts were built on the Mackenzie and the beginnings of the fur trade in that rich empire were set in motion. Itself isolated from the main scenes of conflict in the courts of Montreal, on the westward trails, and in the carnage of Red River, but with local alarums and bloodshed of its own, the Mackenzie northwest was none the less greatly affected by the impact of the fur war. Now the white man on the river, the Indian middleman, and the native trapper could buckle down to the business that had brought the first two into the country of the third, the catching and marketing of fur. The years that followed, from 1821 to 1869, were the golden era of the fur trade in the empire of the Mackenzie.

CHAPTER SIX

The Golden Years

MERGER of the two principal competitors for the fur of the northwestern empire did not necessarily mean the end of hostilities on the far-flung outposts of the trade, for communications in 1821 were a slow and arduous business, virtually suspended in winter and involved with time and the rivers during the brief summer season. Thus, throughout the north, embattled traders continued to fight each other tooth and claw, in some cases for months after the two companies had become one, simply because they did not know of the peace. In the vast basin of the Mackenzie and its tributary waters, war continued long after agreement had been reached, particularly in the vicinity of the Nor' Westers'

Fort Chipewyan on Lake Athabaska, headquarters of their Mackenzie trade, and nearby Fort Wedderburn, the base of the Hudson's Bay Company.

To Lake Athabaska in 1820, a year before the peace, had come a young man who would become the greatest single figure in the history of the trade down north. He was George Simpson, fresh out of the Bay Company's countinghouse in London, whence he came under sponsorship of Neil Colville, a Member of the Committee of the company and, later, its governor. William Williams, the company's northern governor at the time, lived in daily anticipation of a warrant for his arrest from Montreal, whither the Nor' Westers proposed to bring him to face charges in their own courts. What was needed by the Bay Company in the fur country, therefore, was a replacement for the senior man in case he should be ambushed and spirited away. Hence Simpson was sent to join Williams at Norway House, on the eastern side of the height of land that separates the Mackenzie empire from the rivers that flow down to Hudson Bay. While Williams waited on events he sent young Simpson over the rivers and the 12-mile La Loche Portage, main line into the Athabaska country from the east, to take command of company operations in the area.

On July 30, 1820, Simpson led a brigade of fifteen canoes up the Saskatchewan, over the long portage to the headwaters of the Mackenzie system and on to Lake Athabaska and Fort Wedderburn, a mile from the Nor' Westers' headquarters at Chipewyan. That was the last of the fighting winters, but by all accounts it was a hot one. Competitive traders ambushed one another as they traveled over the ice along the river and in the bays of the lake. Men were maimed. Loads of furs were captured; stolen would be a better word, but it was a two-way piracy. Simpson's neighbors at Chipewyan were led by Samuel Black, notoriously one of the most reckless and daring men in the country, soon himself to be a member of the company he now fought. But for all Black's aggressive reputation, young Simpson held his own, with the result that when the ice went out in the spring of 1821 he brought his brigade of canoes back over the portage and down to Norway House, gunwale-deep in bales of fur, collected from as far away as the banks of the Peace and convoyed to Wedderburn under the noses of Black and the Nor' Westers by armed parties. He arrived at Norway House to learn that peace had been signed many weeks before and that during the closing weeks of battle around Athabaska the principals in the running fight actually had been partners in the same business.

A sharp change now came over the face of the trade in the Mackenzie country. The days of dog eat dog were over, but what they left behind would not be repaired and the house set in profitable order for long months to come. The Indians had grown used to competitive bidding for their fur; now they had a monopoly to deal with, and the first duty the monopolists set for themselves was to cut trade rates and build up profits to replace the lean returns of the long years of warfare. The Indians turned sulky and, in many areas, refused to go out on their trap lines. The new company, meanwhile, set about starving them into submission, a policy which, in the Mackenzie country, soon brought results. Boatmen and roustabouts, who had drawn high wages in the days of all-out competition, now found themselves faced with the necessity of taking lower wages or quitting the country, which left most of them with no option – and no esprit de corps. Deciding to use the Hudson Bay route and to work up the rivers from the coast, the company jettisoned the overland route from Montreal via Lake Superior as a trading highway into the northwest, with the result that men who had earned their livelihood at points along the way, including many who had settled at Grand Portage on Superior, main transfer point of the continental journey, suddenly found themselves without employment and with little, if any, means with which to leave the region. The monopolists were cracking down. The trade was leaving Canada and returning to England where, in the opinion of the fur barons, it had always belonged.

In September of the first year of peace (1821) young Simpson was appointed governor of the Northern Division of the reorganized company, a title tantamount to that of emperor of the Mackenzie. Alone and undisputed, he ruled a country almost a million square miles in extent, throughout which his word was the law. And he ruled it with an iron hand. Gone were the days of the reckless trader, the devil-take-the-hindmost and the hell-with-the-expense trafficker in furs. Now the era of the man of business had dawned, and George Simpson drove for business and profit in his wilderness empire in precisely the manner of a man directing an urban trade from a countinghouse in Montreal or London.

The first task was to rid the Company of recalcitrant Nor' Westers and to incorporate old enemies of reasonable views into the new organization, amongst the latter Samuel Black of Chipewyan, of whom Simpson wrote, long afterward, apropos their first "peaceable" meeting: "Black could, at first, hardly look me in the face. He remembered my Athabaska campaign and never will he forget the terrors in which he was kept that winter. We parted excellent friends." Nor would Simpson ever forget that first winter in the wilderness, for of it he wrote: "On two or three occasions I went for two or three whole days and nights without having a single morsel to swallow."

But now times had changed. Simpson set out to get to know his domain, to visit its forts and outposts, learn to know its Indian peoples and study the differences in the characteristics of Slaveys, Beavers, Dogribs, and Hares from those whom he had known on other waters. Other officers, "active and discreet," he sent to explore the tributary rivers of the Mackenzie itself, not in search of any passage to the Indies, but for new areas out of which he could bring fur to forts to be built wherever trade could be developed; fur to be counted and baled and loaded into the company's ships at the Hudson Bay coast, to replace the casks of rum and brandy, the firearms, the metal animal traps, the strange assortment of trade goods brought out from England to exchange with the natives for their priceless merchandise.

Simpson was no Indian lover – but what white trader ever was? Of relations at the time of amalgamation, Simpson has written: "I have made it my study to examine the nature and character of Indians and however repugnant it may be to our feelings, I am convinced they must be ruled with a rod of iron to bring up and keep them in a proper state of subordination, and the most certain way to effect this is by letting them feel their dependence on us. . . . In the woods and northern barren grounds this measure ought to be pursued rigidly next year, as if they do not improve, no credit, not so much as a load of ammunition [should be] given them until they exhibit an inclination to renew their habits of industry. In the plains, however, this system will not do, as they can live independent of us, and by withholding ammunition, tobacco and spirits, the staple articles of trade, for one year, they will recover the use of their bows and spears, and lose sight of their smoking and drinking habits; it will therefore be necessary to bring those tribes round by mild and cautious measures which may soon be effected."

Here spoke not the reckless frontier buster but the conservative and tough man of affairs. Turn down the screws on the Indian where you can get away with it; take it easy where you might stir up trouble to hurt your own trade. The golden days were over for the northern Indian, but they were just dawning for the traders, with their newly gained monopoly in the north.

The amalgamation, as noted, brought about a complete change in transportation routes into the country. Now ships discharged their cargoes for the Mackenzie at York Factory on Hudson Bay, whence goods and supplies came up the rivers on the far side of the watershed until they reached the 12-mile Portage La Loche over the height of land, from the summit of which the loads went downhill to the banks of the Clearwater, which joins the Athabaska at the site of Fort McMurray, the present end of steel. Thence the freights were rowed, or sailed, down the Athabaska, through the delta and into the lake, to be unloaded at Chipewyan, for breaking down and reloading, before moving north to the forts on the Mackenzie, as far as McPherson, far beyond the Arctic Circle, the Lower Ramparts and Arctic Red River, on the fringe of the Mackenzie Delta, more than a thousand crowflight miles from "Chip," and an over-all water journey from the sea of almost two thousand. The end of the long water trail brought them back to the marge of tidewater.

Now a new type of transport had been brought into the country, suitably named the York boat, from the company distributing post of the same name. The new boat, which replaced the 30-foot canoes of Mackenzie's day on the river, was a shallow-draft craft about forty feet in length, with 10-foot beam, shaped like a whaleboat, with sharply angled stern and bow. It carried oars and a sail, but was usually "tracked" upstream, the crew hauling on a hawser as they made their way along the shore. The new boats, built in large numbers at Chipewyan, were eight-oared and steered by a ninth, handled by the senior voyageur standing in the stern. By comparison with the canoe, these craft would carry great loads, as much as two tons, in addition to crew and crew supplies, and were infinitely more seaworthy. Their introduction by Simpson in 1826 typified the new businesslike approach to the days of peace.

The York boats traveled in brigades of a dozen or more. Wintering at Fort Simpson, the Mackenzie brigade would leave the mouth of the Liard as soon as the ice began to move out, following it down to McPherson, where the season's catch of fur would be loaded and summer supplies and trade goods left behind. Then began the long upriver journey to Good Hope, Norman, and other river forts to pick up the winter's fur. Gunwale deep in bales of marten, beaver, mink and ermine, the boats pushed back to Great Slave Lake and over Alexander Mackenzie's route to Chipewyan. From Chipewyan other brigades carried the catch as far as the long portage at the Clearwater, where crews of roustabouts manhandled it over the height of land. On the far side the bales were turned over to other York-boat voyageurs for transport to York Factory, there to be loaded into seagoing ships. Viewing the process in reverse, ships which left Britain in the spring would unload their trade goods at York Factory in August, too late to be carried into the country during the same open season. Cargoes were sorted and stored at York Factory during the winter, moving forward after the ice left the rivers in the following June. Only by a well-timed operation, in which the brigades of boatmen kept to a closely arranged schedule for contact at the portage, could trade goods which had left England in the spring of, say, 1830 be delivered to distant forts on the Mackenzie before the rivers were blocked by ice in the autumn of 1831.

While the old Nor' Westers may have believed that they had won the better terms in the amalgamation, the fact remains that with the coming of George Simpson the new monopoly immediately turned back to the practices of the Hudson's Bay Company. Headquarters were again in London. Montreal quickly ceased to be a center of the traffic in fur and Canadian personnel began to disappear from clerical and managerial posts in the north, as quickly as old-timers died or reached retirement age. The major part of the northern office and executive staff was recruited in Britain, principally from Scotland, if only because it was easier to bring men across the Atlantic by boat than overland from Montreal and also, it may be suspected, because the Orkneymen, who came out again in large numbers, being accustomed to British rates of pay, would work for less money than young Canadians.

With the coming of peace, the fur trade became big business in the wilderness, as tightly organized under Simpson as any urban enterprise. Communications were speeded up: in summer by introducing fast, express service on the river route; in winter with dog teams on the frozen waterways. The record of a typical winter mail-and-dispatch delivery, hauled by a team of seven husky dogs, is noted in an old record. It reports departure from Fort Simpson before daybreak on January 22 and arrival at Norman late in the afternoon of February 6, a journey of more than three hundred miles, the driver noting casually in his report that he had been delayed by snow. Next morning he was off at daybreak over the frozen surface of the Mackenzie, pointing for Fort Good Hope, which he reached on the sixth day out. At daybreak the driver was on the trail again, heading for the distant Yukon. Today the air voyageur may take his breakfast in a steam-heated company hotel, near the oil refinery at Norman, and be far beyond Simpson, on Great Slave Lake, and Fort Resolution, hard by the mouth of the Slave, in time for an early lunch. But in the golden years Simpson's "express" was regarded as a speed-up of revolutionary proportions, typical of the efficient trader who inaugurated it.

Throughout that part of the record of the fur wars which has been written by loyal Bay Company men, the irritation and disgust of the British company with the ways and techniques of the get-rich-quick Montreal traders may be constantly read in and between the lines. They deplored the Canadian's free-flowing use of rum, not out of love for the Indian nor concern for his morals but in belief that rum would debauch the native. Yet the lords of the Bay Company could not themselves escape making rum available in quantity even long after the amalgamation, for the Indian who had done business in the days of all-out competition would trade with no one who did not offer spirits as the principal medium of exchange.

Often scenes of the wildest debauchery were witnessed at the frontier forts when tribesmen came in with the season's catch. Barring a few necessities to sustain life, items of food, new trapping equipment, the remainder of payment (itself ridiculously low in comparison with the value of furs in the European market) would almost invariably be taken in spirits and tobacco. Records of the early days tell of native orgies in the vicinity of fur-trade posts in which not only the braves, but squaws and adolescents, alternately drank, fought, slept and drank again, until the value of the season's catch had been spent. Only then would the Indian take himself and his hangover back to the bush, there to resume his role as trapper until such time as he had another consignment of fur ready to be taken to the fort to be traded for rum.

The spirits problem, then, was one of the first the Bay Company tackled, once more in supreme control of its empire. The reason for the tightening-up-on-liquor policy was crystal clear. The Indian was the company's most valuable instrument and the source of its profits. Without him, who would bring in the fur? Liquor he would never learn to "handle," perhaps because he had no background or knowledge of spirits to support him, but had been introduced to alcohol by the white man after centuries of abstinence, perhaps because his personal chemistry and emotional balances could not be adjusted to its use. The fact remained that if enough rum was traded to enough Indians over a sufficient period, in due course there wouldn't be many good trappers left – and what the Company, with its long-range outlook, realized was that its whole future was rooted in the native who trapped the fur and brought it to the post. For this reason primarily, and certainly as no great humanitarian project, the Bay Company under George Simpson set about cutting the Indian's rum ration, until spirits finally disappeared completely from the post manager's stock as an item of trade.

But this was by no means the only change in the processes of getting fur out of the wilderness and over its complicated lines of communication to market. In challenging those who have said that it was George Simpson's personality which revolutionized the northwestern empire, it could be said that Simpson was simply giving effect to a policy which had never changed since the Honorable Company had been chartered in 1670. It was a policy designed for monopoly (which the company legally possessed), but when strong competition entered the company's domain that policy had to be shelved and the competition met on the interlopers' terms. But the policy itself was ironclad. Now Simpson was returning to the policy.

It was rooted, as noted, in the desirability of monopoly. As the owners of the company in Britain saw it, their vast arctic and subarctic domain was in fact a nation, owing allegiance to the British crown and to nobody else. Within its domain the company's wishes were law and the company, for its own benefit, provided whatever machinery of enforcement might be required. It was extremely simple and rugged machinery, reasonably tolerant and just, but designed to turn over for the benefit of the handful of whites who directed the company's affairs in the north and their principals across the ocean. It has been called a benevolent dictatorship, but there were occasions when the reasons for use of the qualifying adjective could not be easily discovered. The purpose certainly was not to raise the standards of living or the literacy level of the native, nor to indulge in broad humanitarian projects. There are plenty of items in the record to attest to lack of interest in the education of the native or efforts to Christianize him. A good Indian was an Indian who brought in a lot of fur and who depended solely and entirely on the company for such livelihood as it chose to give him in return. A bad Indian was one who didn't. Educate him and God knows what he might do or want next. It was as simple as that.

For all its rigid approach to its purpose in life, however, much may be said for the policy of the company, vis-à-vis that of its earlier competitors. The latter were get-rich-quickers no more concerned with permanence than with the welfare of the Indian. If one section of the country should be milked dry of fur-bearing animals, or the native population became debauched or unmanageable, you simply moved on, for the fur empire, as the Canadian traders saw it, was as limitless as the northern horizon. Not so the Bay Company. The Adventurers of England trading into Hudson Bay were men who regarded their concession from the crown as establishing a permanent empire and enterprise, with accompanying responsibility to treat it well. Hence they soon put into effect policies of fur conservation and replenishment. They looked to their tangled skein of water communications as lines to be guarded, tended, and improved. Their transportation experts were always on the lookout for ways to cut long journeys over regularly plied routes by a day, two days or three days, for days saved meant money earned. Above all, they brought into the country organization and efficiency of operation, in perhaps the most difficult corner of the world in which these have ever been put to use. No Johnnie-come-latelies, these Gentlemen of the Bay. Whatever interest in the common good they may have lacked (and nobody was giving much thought to the common good on the frontiers of the world a century ago), at least they brought dignity and order to the country, qualities completely missing during the hobbledehoy days of the wars with the Montreal traders.

Undoubtedly the Montreal traders' methods had bewildered the men of the Bay, accustomed, even to their factors and traders in the field, to the commercial practices of the British Isles. What angered them perhaps even more than the Montrealers' methods was the fact that they themselves had been forced to stoop to similar practices (as in the overliberal use of spirits in trade with the Indians) to meet the Canadians' competition. But now the men of commerce, the penny-pinchers, the conservatives had won out, thanks to the royal charter which created the monopoly and gave its holders an empire to rule. At first perhaps it seemed that the Nor' Westers had won. But where was the Montreal fur trade? It had virtually ceased to exist. Where was the new center of its direction? It had recrossed the ocean to London. Who were the headmen in the field and to whom did they give loyalty? Within a few years most of the Canadians had been shaken out (and there is ample evidence of the shaking-out process, barring a few who could be fitted into the new picture). Many of those who stayed could not accommodate themselves to the Orkneymen traders, whom they disliked by instinct and who, in return, made no attempt to hide their contempt for Canadian trading ways. When a Canadian moved along, a Scot almost invariably came out to take his place. In short the Old Company, with its dignity and its tough-minded decorum, had won the fur war. That fact comes through the record of the times with a distinct and undeniable clarity. This was the Bay's empire again, and Bay empire it would remain so long as its monopolistic rights endured.

Thus it was not the personality of George Simpson that was responsible for the golden era of the fur lords, but the lords themselves and the policy they had long since laid down. This is not said to detract from Simpson's tremendous capacities as organizer and ruler, for they were gigantic. But Simpson stemmed from a policy and he learned it himself in the days when, as a youth, he worked in the London countinghouse of the company, until it had become embedded in his personality. Hence Simpson had come to Canada with his thinking deeply rooted in a policy that was almost a religion. Simpson was an instrument, a splendid instrument, but not a policy maker.

Simpson lived as befits the ruler of an empire, not because he wanted soft living, for he seems to have been a frugal man, but because that was his role as policy saw it. When he moved about his vast domain he traveled as a monarch, in so far as the circumstances of the country permitted. He rated gun salutes and flag ceremonies as he approached a post or departed from it. Special quarters were prepared against his coming. Banquets were laid on for his delectation. When he spoke, his words were law, to be obeyed without question by all of inferior rank. The chain of responsibility reached down through the ranks in similar terms. Within his own orbit, every satrap in the field, the chief factor, the chief trader, the manager of an outpost, held supreme command and within that command he was what might be called a reasonable facsimile of God. Discipline on the ladder of seniority was as rigid as in any army. Officers ruled those beneath them with what Simpson himself called the rod of iron. Each gave to the man immediately above the unswerving obedience he himself demanded from those below. Men in positions of trust and command were commissioned as in an army and did not become "gentlemen" until they had a promulgated commission to prove it. The company's Standing Rules and Regulations covered sixteen printed pages and were military in style and concept. The domain of the Bay Company was, in fact, a quasi-military state. That this system worked far better than the catch-as-catch-can methods of the Nor' Westers and others of the ragamuffin days, is clearly established by the record of the half century during which the men of the Bay were in supreme and unchallenged command in the vast empire of the Mackenzie.

From 1809 to 1814, when the fur war was at its height, no dividends were earned or paid. From that time until the year of amalgamation, the committee in London managed to scratch together a payment of 4 per cent, at which rate it remained for three years after the coming of peace and monopoly in 1821. Those were the years of reorganization. Then earnings began to jump. In 1825-26-27 the dividend rate was 10 per cent. The next year it was doubled, and fluctuated between 20 per cent and slightly lower figures until 1836, when the owners received a 23 per cent return on their investment and shares were quoted in London at two and a half times the price at which they were traded in the first year of peace. In 1838 shareholders were receiving a flat 25 per cent in profits per annum, the peak year of monopoly prosperity. Thereafter trade began to flatten out and in the 1840's and 1850's the annual dividend ran at 10 per cent, around which figure it remained until rumors of confederation between the Canadian provinces were heard in the 1860's and political gossip indicated that the new Dominion would insist on taking over the Bay empire, permitting the company to retain ordinary trading rights, but insisting on surrender of its charter and on the restoration of competition in the fur trade. By then huge fortunes had been amassed in Britain from the monopolistic methods of the company in Canada, a trade in which Canadians had practically had no hand for almost half a century, a trade which handled virtually no Canadian goods, because of the direct sea route from the Hudson Bay coast to Britain, and made no use of Canadian facilities for dressing, marketing, or shipping fur. In Canadian opinion the company was simply removing one of the country's richest resources without compensation. The view of the company, of course, completely reversed that of Canadians. To the lords of the Bay Company and their satraps in the far-flung fur forts of the Mackenzie, the north was a country in itself. If it had cut itself almost completely off from Canada after the amalgamation, by reversing its lines of communication and leading them south from the Arctic coast to its inland stations, it did so by right, for as the Bay barons saw themselves, they were Britishers, not Canadians. To the men of the Bay the Canadian was still an interloper, a potential poacher, the north an empire in its own right.

That monopoly brought increase in the output of fur is seen in the figures of the period. In 1821, the first year of what Simpson liked to call the "coalition," the Mackenzie country supplied only 111 packs of fur. Ten years later production had gone up by one-third. In 1852 the Liard River posts alone put up 32 packs, an estimate of the size of the pack or bale being found in the figures of 4,200 marten skins, 1,582 beaver, and 85 bear in that season's Liard catch. As this represented less than one-fifth of the whole Mackenzie export for the season, the extent of the traffic becomes measurable. In the years 1853-54-55, the Mackenzie River posts sent out 137,132 marten alone, a remarkable figure when considered beside the difficult circumstances under which the trade had to be carried on. Monopoly, bringing discipline to the now-dependent native and order to the country, had raised production and reduced costs in remarkable degree.

Simpson's Bay Company factors drove up the tributaries of the Mackenzie and far into the western mountains in their search for new areas of profitable trade. Posts were built, opened, but quickly abandoned if they failed to appear on the right side of the ledger.

In 1856, when the monopoly trade could be said to be close to its peak, eleven posts were in operation on the great river and in its tributary areas, with headquarters at Fort Simpson, situated on an island at the junction of the Mackenzie with the Liard. On the Mackenzie itself was the post on Big Island, at the point of outlet from Great Slave Lake, used for many years not merely as a trading establishment but for wintering quarters for staff working in the upper river areas. Good Hope and Norman were the principal stations on the river north of Simpson. Fort McPherson on Peel's River, which flows into the Mackenzie delta, was built in the second decade of the monopoly and in 1846 a company trader named John Bell, who had built it, moved back inland to become the white discoverer of the Yukon, where the company immediately built Fort Yukon, which it occupied for twenty-two years, when settlement of the

Alaska Boundary dispute forced the traders out, as its establishment was on what then became recognized as United States territory. Rampart House was built near-by on British soil but was soon abandoned, as the trade there was unprofitable. Pierre's House, deep in the Rockies, suffered a similar fate, as the Gentlemen Adventurers learned that the greatest gain was to be had along the great river and its immediate tributaries rather than in inaccessible country. On the Liard itself were Forts Liard and Halkett, and later a smaller station, known as Toad River, was built midway between the two main establishments, but subsequently abandoned. Sometimes posts were moved to meet the vagaries of trade and climate. Fort Norman's original site was at the mouth of the Bear River. Later it was moved 20 miles north, then back again, then moved for a third time. But in time a greater sense of permanence came to the posts on the great river, operating under direction of a chief factor stationed at Simpson. Big Island was virtually abandoned and Fort Providence, 40 miles down from Great Slave Lake, became the main post of the area. Simpson itself, then Norman, Good Hope and McPherson, the last-named built for trading with the coastal Eskimos, became, with Fort Providence, the permanent Mackenzie stations. Later Arctic Red River was added, south of McPherson, and finally Wrigley on the lower reaches of the river, midway between Simpson and Norman completed the picture, as it does to this day.

The white man's life in the river forts during the golden period was arduous, but not without its compensations. The governor, as director of broad policy, saw to it that the main posts were rendered as comfortable as the circumstances of the country would permit. The post itself would usually occupy a clearing of several acres, closed in by a stockade if situated in country where defense might conceivably be required. The buildings were originally constructed of logs, some, later, of dressed lumber, ferried down the river. Houses were sealed against the elements, though they still were extremely difficult to heat in a climate in which temperatures touch as low as 65° below zero and the mercury in the thermometer ceases to function. Double windows and doors kept out most of the weather, but not all of it, as witness reports of buckets of water adjacent to the door freezing, in a room in which the stove in the center glowed red with heat.

In low-lying areas the buildings of a fort, as on the banks of the Mackenzie, were raised well above ground, as protection against the floods that gorged all the rivers when the ice began to move, jamming in the narrow bends and backing up the great volume of water flowing down from Great Slave Lake and the tributary rivers. Inside, buildings were commodious and comfortable. That of a chief factor might even verge on the sumptuous, as befitted the status of the ruler of a province or principality. Married men's quarters and the bachelors' hall might lack fancy trimmings, but were comfortable and, so far as the weather would allow, warm. Larger establishments which served as distribution points as well as trading posts, such as Fort Simpson as headquarters for the Mackenzie, would, in addition to staff concerned solely with the local trade, include a surgeon, an accountant (charged with auditing the books of a whole district), and a postmaster. Self-containment was the necessity and fact of trading-post life on the river in the days following amalgamation.

During the winter, which occupies eight months of the year, weeks might pass without arrival of a visitor, excepting the occasional Indian trapper come to trade his fur and acquire supplies. Until Simpson inaugurated his express dog-team service, long months passed without so much as a letter arriving, but once the dog-team mail had been established, two or three mails could be expected during the long arctic night which, reversing the 20-hour days of summer, brings down an equal period of darkness and dusk.

The brief summer comes suddenly. Today's winter seems almost to become tomorrow's spring. Sap rises in the shrubs and bushes, buds burst into greenery. Days lengthen until the almost continuous presence of light and sun produce a growing season which, in terms of speed, is unrivaled on the continent. Soon the white traders on the river were clearing ground outside the actual living area and planting gardens, an activity typical of the new policy, as opposed to that of fur-war days, when any self-respecting trader would have considered such bucolic notions as beneath him. Now posts were encouraged to grow green vegetables wherever soil and climate permitted, again for the utilitarian reason that whatever bulk could be produced on the spot meant less to be carried down the rivers from Chipewyan by the York-boat brigades.

Summer meant not merely relief from the isolation of the winter, but an even more welcome change in diet. Traders and clerks ventured into the surrounding country to shoot ptarmigan or wood partridge. Fresh-caught fish appeared on the table. As the white men foraged for fresh meat, Indian and half-breed helpers turned to hunting. During the migrations of ducks and geese hundreds were shot and "laid down" for the next winter, when they would appear on the table, dried. The spring brought the northward movement of the great caribou herds, providing another source of fresh meat as they passed by and another laying away of food for storage.

Until the posts, under Simpson's businessman regime, turned to truck gardening, no fresh vegetables had ever been seen on far northern tables and until the arts of canning were developed "outside" and the conservative gentlemen of the Bay felt they could regard this method of food preservation as dependable, not just an innovation, meat and fish were the staples of the fur trader's twelve months in the year diet, and by all accounts they were trenchermen of the first order. Time and again in old records discussion of the food which weighted down the board indicates that for a man to put himself on the outside of a couple of ducks or geese at dinner was considered ordinary, for mammoth stoking of the human interior was regarded as necessary to men leading vigorous lives in a strenuous climate. The furmen thrived on it, no matter with what horror the modern dietician would regard their daily menus, perhaps because the Scots are a hardy race or maybe for the simple reason that theirs was a simple, healthy life.

Truthfully it may be said of the Mackenzie empire that it has changed little in more than one and a half centuries since white men first moved in. The areas of recent natural resource development have tended to become modern communities, addicted to electric light, running water, indoor plumbing, and the more effete life of southern Canada. Not so the isolated fur posts still in operation along the banks of the Mackenzie. More than one enthusiast has dilated at length on the manner in which the coming of the airplane has revolutionized the business of living in the Far North. It has done so but not in precisely the manner that might be imagined. Today the country is dotted with airstrips and mail planes call at Simpson and Wrigley, Norman, Good Hope and McPherson and wing on over the delta to Aklavik. Sometimes they fly in fresh meats and vegetables. Always they bring newspapers, magazines, all manner of things commonly called amenities. But in the great gaps between communities, the dog team and the trapper's canoe (equipped today with outboard motor) are still the primary means of transport along the lifeline of the country. The great river remains the highway, not the air above it but the water itself, liquid or frozen. The river is still the artery of the fur trade and so it will be so long as the fur-bearing animals remain and human beings covet possession of their silken hair.

There were giants in those days, men to match such earlier giants as Hearne, Dease, Rae and Thomas Simpson, the relative of the governor who had cut so wide a swath in the annals of northern exploration. Some of those giants were men of strange character and peculiarities. There was John Rowand, who, though never actually a trader on the Mackenzie, for many years operated the great post where the city of Edmonton (present aerial gateway to the Territories) now stands. Fort Edmonton was more than a trading post. It boasted shipyards for the mass production of York boats. It served as a packing house for dried-meat pemmican packed in leather sacks and shipped north in quantity to be consumed by the boatmen of the Mackenzie brigades. Wheat, barley, and vegetables were raised in a large tilled area near-by. The fort, a place of studded palisades and flying pennons, boasted, in addition, the Great House, built by Rowand, a reflection of himself and the ruler-mindedness of the senior men of the trade.

The house stood 60 feet by 70 and three stories high, its large windows of "real" glass and not of the skin parchment widely used in the interior. On one side of the main entrance was a comfortable mess for the commissioned gentlemen, on the other a huge room where the chief trader held court for visiting Indian celebrities and tribesmen. Of this room a visitor once noted: "The walls and ceilings are boarded, as plaster is not used, there being no limestone within reach, but these boards are painted in a style of the most startling, barbaric gaudiness and the ceiling filled with centre pieces of fantastic gilt scrolls, making altogether a saloon which no white man would enter without a start and which the Indians look upon with awe and wonder."

Here Rowand presided over native feasts and dancing. The combination of personal courage and the magnificent gestures of which he was capable filled the Blackfoot Indians of the immediate region with an awe equal to that inspired by the ceremonial room at the fort. Old records speak of such feats as his singlehanded dispersal of Indian raids while on the trail. Once he is said to have walked out to meet a party of two hundred Blackfeet in full war paint as they descended on his encamped party, the Indians obviously out for scalps. Rowand, says the legend, went out alone, raised his hand high and called, "Stop, you villains!" The villains stopped, abject in their apologies when they discovered that they had been about to attack the camp of the little man from Fort Edmonton. A myth? Perhaps. Such men under such conditions create their own mythology.

When Rowand died at his fort and his will was read, it divulged his desire not to be buried in the Indian country, but among his own people, in Lower Canada, and what happened in fulfillment of the trader's last wish is a matter of record. Governor Simpson ordered that Rowand's remains be pickled in a cask of rum and carried in his own canoe to the Red River settlement. Rather than run the risk of having the keg and Rowand dropped overboard by superstitious York-boat voyageurs on the southern route to the Great Lakes and Montreal, Simpson sent the body north from Red River to the Hudson Bay coast at York Factory, whence it was taken to England with the next season's boat and shipped back to Canada for final burial. So John Rowand went home to his last resting place, spending more than a year on the way, in a cask of fur traders' rum.

There was Robert Campbell, who, sent to the Mackenzie in 1833, spent seventeen continuous years in opening new country to the trade. Campbell was not the Rowand type, excepting in one quality, courage. He was dour and taciturn, given to Bible reading and to long hours with his nose deep in Hervey's *Meditations*, which apparently had the power to remove him from the severe conditions of life about him. In venturing into unknown country, time and again he met native bands, more warlike than the docile tribes of the Mackenzie, who faced him with bows drawn, ready for the kill. When that happened, Campbell would put down his own weapons and raise his arms high in the sign of peace, which never failed him. It was Campbell who explored the treacherous Liard which, by 1852, had taken the lives of fourteen fur traders. High up its waters, in a forested valley, he came upon a great Indian celebration, presided over by a native queen, described by Campbell as a "woman of superior intelligence." Although her braves resented this white intrusion, the queen welcomed the trader, who remained as her guest throughout the fete.

It was Campbell who plunged through the mountains in 1851 to discover the Pelly River, following it through gorges and over cataracts, to discover that what he had found was the upper fork of the Yukon, for his river brought him downstream – part way on a handmade raft, the rest of the journey in a boat built on the bank without tools – to Fort Yukon, erected four years earlier, by Alexander Hunter Murray.

Murray was a Scot, out of Argyllshire, who had served an apprenticeship with the Astors' American Fur Company before going down north. On his way into the country he met and fell in love with a daughter of Chief Trader Colin Campbell at Lake Athabaska and they were wed by contract by another chief trader, Murdoch McPherson, by a simple written and spoken pledge, which the Canadian courts recognized as binding when taken in a country without clergy.

Bride and groom spent their honeymoon traveling by York boat down the Mackenzie to the edge of the delta and Fort McPherson, where they wintered. There Murray left her in the spring, driving west into what was then regarded as Russian territory, to reach the Yukon River and build the fort that bears its name. The fort he built was one of the most defensible in the country. It needed to be, for Murray expected the Russians to come up from the sea to expel him. The pickets of his palisade were full-grown trees, peeled and squared to fit, rising 14½ feet aboveground and driven more than 3 feet into the earth. "When all this is finished," Murray wrote, "the Russians may advance when they damn please!"

No Russians came. Instead, the Indians began to arrive at the fort in great numbers, bent on the peaceable pursuits of trade. Fort Yukon, most remote of all the company's establishments, remained a profitable center until settlement of the Alaska Boundary dispute forced its owners to move on. So far was it from the marts of trade that a period of seven years elapsed between the time of departure of trade goods from England and the arrival in London of the furs for which this merchandise was exchanged deep in the mountains of the arctic. A company report lugubriously describes this as "a long strain on invested capital," as always the hard core of policy.

There were other giants. The history of the country is alive with their names. Forts, lakes and rivers, almost without number, have immortalized them in letters on the map. There were giants in those days.

But the golden days were soon to end. To the south and east Canada began to stir with the growing pains of nationhood. Long-separated provinces began to put aside petty differences to discuss a common threat, the threat of piecemeal annexation to the United States, if they failed to come together and create a country of their own. While Canadians looked south and pondered the problem posed by a not-always-benevolent Uncle Sam, they occasionally turned their eyes north to ponder another: how to break the fur monopoly and incorporate into the country the vast region which the northern barons insisted was their own, an empire giving fealty to Britain, and, they said, beyond Canadian intervention.

Perhaps the fur lords had counted on the support of the British government for their view. If so, they had neglected to note that statesmen across the sea were themselves turning away from the old imperialist concept of concessions and monopolies in the far-flung corners of the Empire.

In 1867 the Canadian provinces confederated and became a nation. Two years later the golden years of monopoly ended, when Canada took over title to its own territory from the company, smashed the monopoly, and reopened the north to competitive trade. In fifty years of peace the fur barons had done well in some degree by the country, to which they had brought law and order, and certainly more than well by themselves.

CHAPTER SEVEN

A Free Country

IN THE EARLY DAYS of the white man's rule in the empire of the Mackenzie, the fur traders had lived almost entirely on the wild life of the country. Later as the organization of the Company became more elaborate and food-preserving techniques improved, greater quantities of prepared food were taken into the interior. At the same time agricultural bases were established to supply foodstuffs for the north. All this was part of Governor Simpson's businessman policy, and from the viewpoint of the trade and the company it was excellent, first because it removed the uncertainty of living, for without a backlog of supply a change in the migration habits of the roving herds might threaten an isolated post with starvation. Moreover, money was saved by removing the necessity of bringing prepared foods by sea from Britain.

But the opening of agricultural areas on the southern prairies had another and unexpected impact. Around these agricultural bases the population of the agrarian west began to spring up. First came former fur traders, their Indian wives, and half-breed children to establish permanent settlements, principally around the Red River area. The coming of the Selkirk colonists established a white population completely independent of the fur trade. Finally, the drastic reductions in personnel that followed amalgamation of the companies confronted many surplus employees with the problem of a livelihood, with the result that they settled in the prairie area and became farmers. None of this had the Bay Company foreseen. By the 1860's a considerable farming population had settled in the country, where it lived under the political rule of the Hudson's Bay Company and became increasingly restive as the years passed.

This population was not content to allow its interests to be subordinated to the fur monopoly, and its leaders did not hesitate to say so. Company officials continually (and often correctly) suspected the farming population of illegal trade in furs, and the measures adopted to combat this were extremely unpopular. Protests were made by the half-breeds in which they claimed that they could not enjoy their full rights as British subjects because of company policy. In 1844 the governor issued two proclamations. One required that "each settler, before the Company would carry any goods for him, should be requited to declare that he had not been engaged in the fur trade." The other insisted that "the writer of every letter write his name on the outside of it, in order that, should he be suspected of dealing in furs, it might be opened and examined." Another device was imposition by the company of a 20 per cent tax on imports into its territory, even from Canada, but exempted those who were not suspected of dealing in furs. Besides this a standard land deed was drawn up, to be signed by any who wished to hold land in any part of what is now the Canadian west. Under the terms of such a deed, any landholder who encroached on company privilege forfeited his lands. Finally, since practically all transportation was in the hands of the company it could name its own freight rates, which it did, often solely for the specific purpose of checking trade.

Such a policy was bound to result in protests on the part of the settler. In 1847 a young London barrister, part Canadian Indian and a former company man, named Isbister, forwarded a petition to the Secretary of State for the Colonies, which charged the company with growing rich at the natives' expense, retarding their progress, and neglecting their religious conversion. These charges created a disturbance in the Colonial administration in Britain and the company was forced to explain itself. This it did, adequately so far as the British government was concerned, and nothing was done.

At the same time as Isbister's petition, another was presented signed by 977 French half-breeds. They asked: (1)

that as good subjects they might be governed by principles of the British Constitution; (2) that as British subjects they be granted their right to enjoy the liberty of commerce; (3) that land be made available to newcomers, a portion of the proceeds from its sale to be applied to improve the means of transport in the country.

The French and English half-breeds united in bringing pressure on the governments of the Canadian provinces and on the Imperial Parliament in London. Agitation kept the populace restive and frequent incidents of insubordination against company rules and of mob violence occurred. The founding of a newspaper, the *Nor'-Wester*, at Fort Garry on the Red River, did much to add fuel to the flames. By 1868 it was clear that the company had lost its political grip.

The legal monopoly and political control by the Hudson's Bay Company, on territory outside Rupert's Land, was based on a charter renewable every twenty-one years, a license scheduled to expire in 1859. When the Bay applied for renewal, the Colonial office decided to hold a general review of affairs in British North America, and invited the Canadian provincial governments to present their views. Canada claimed that its national boundaries should be extended to the Rockies. The company rejected this view. The problem was then placed in the hands of a parliamentary committee known as the Committee of 1857. Ten years after its report was written, the fate of the company was still undetermined. Meanwhile Confederation of the Canadas had taken place (1867) and the new Dominion was in better case to press its claim to sovereignty over the whole of the northern half of North America. Provision was made in the deed of Confederation, the British North America Act, "for the admission of Rupert's Land and the North-West Territory to the Dominion." Canada declared "that in case of Union the legal rights of any corporation, as the Hudson's Bay Company, association, or individual should be respected; that this should be settled judicially or by agreement; that the Indian title [to the country as its original owners] should be legally extinguished; and that an address be made to Her Majesty to this effect."

The Bay barons insisted that an understanding be reached as to terms before giving its consent to transfer, however. Again the question was taken to London, where pressure was put on the company by the British authorities to accept reasonable terms.

The main provisions of the bargain were that the company should surrender all its rights in Rupert's Land, that Canada pay the company the sum of £300,000, that the company be allowed to retain small blocks of land around its posts, and be given one-twentieth of the arable land of the country, and finally, that the company be allowed every privilege in carrying on trade accorded a regular trading company. Monopoly was on the way out. Once more Canadians would have an opportunity to participate in the fur trade.

During the period between passage of the Transfer Act and the taking over of administration of the west by the

Dominion government, considerable confusion existed as to who exercised power in the northwest. Matters were complicated by the illness of the company's governor. In this political vacuum the half-breeds in the west continued to lend willing ears to agitators. Trouble quickly brewed and boiled over.

Under the leadership of Louis Riel, the half-breeds prevented a survey party, sent out by the Canadian government before the Dominion had "taken over," from carrying out its work. They, above all others, had been the hewers of wood and drawers of water through the long years of monopoly, and among the early instigators of agitation for inclusion of the Territories in Canada. But here were men representing the new government coming into their country before they were entitled by law to be there, men without understanding of the point of view of the children of the settlements and with nothing but contempt for the "breeds." The "government men" chose rather to make friends with the so-called Canadian Party, members of the English Protestant minority in the country, whereas the half-breeds (known as the Métis) were Catholic and French-speaking. From the outset of the agitation, though both groups favored inclusion in Canada, bad blood, pouring from racial and religious prejudices, had run between the two.

The scene of the incidents described was Fort Garry, heart of the Red River settlement. Its background was the coming of the surveyors before the act of transfer had taken place. Their manner and deportment immediately convinced

Riel and his Métis followers that if this was to be the way of Canadians in their country, it was apparent that they were simply exchanging one oligarchy for another. This estimate was confirmed when news came through that the new Canadian lieutenant governor, the Honorable William McDougall, was en route into the country from Ottawa to assume his official post, again, Riel insisted, without right or authority to do so, for the "union with Canada" was still not proclaimed. After traveling west by way of Chicago and St. Paul, McDougall was met at the international border by an armed body of Métis who forced him to retire to the American side of the line.

By this time fort and settlement had been taken over by Riel. Equally resentful of the brusque manner in which the Canadians had chosen to move into their former domain, Hudson's Bay leaders on the spot would take no action against the insurrection. To them the whole miserable business was a sellout engineered by their own leaders in London, or at least concurred in as a means to secure high compensation, which, the traders could be sure, would not be divided, even in part, among the executives of the company in the field, but only among the shareholders in Britain. So the traders held aloof from the ruckus, some even giving the appearance of support to the Riel movement.

The half-breed rising had its roots in fear and resentment: resentment against the company for having sold out to another authority which obviously intended to rule them with the same iron hand; fear that Canada would flood the country with Protestant settlers and so quickly install a majority in their largely Catholic land. But at no time did Riel proclaim his rising as rebellion against the crown. It was, he repeatedly insisted, a movement to prevent the lieutenant governor from taking over until such time as terms had been made for the Métis and they could be sure of democratic justice and freedom.

By November 13, 1869, young Riel had taken command at Fort Garry, where he ruled with calmness and restraint. He fed and armed his men from the Bay Company stores and paid his accounts in Hudson's Bay scrip. There was little drunkenness and no violence in the fort. Now Riel sat down to wait the Canadians out. And of the Canadian government it can be said that it bungled the affair at every step.

Meanwhile the government in Ottawa asked Donald Smith, later Lord Strathcona and governor of the company, but then simply regional manager at Montreal, to proceed to Red River and treat for peace. Arrived on the ground, Smith summoned a public meeting at which the people of the settlement elected a committee of forty representatives who, in turn, chose Riel to lead them, thereby establishing provisional government as legal in the state of political vacuum as any other. Riel insisted he would remain "in office" only until terms were reached with Ottawa.

The period that followed, like the earlier days of the "rebellion," was marked by lack of violence on the part of Riel and his people, with one hour of exception. Unfortunately, in that hour a firing squad killed a Canadian Party Protestant named Thomas Scott by Riel's own order, after conviction for looting. The shooting had incendiary results, particularly in faraway Protestant Ontario. Revenge now became the motive of Canadians of the Protestant faith, a spirit superheated in an atmosphere of ignorance and prejudice among people who knew less than nothing of Red River and its problems.

Meanwhile Ottawa had instructed Lieutenant Governor McDougall to proceed to Fort Garry on December 1, 1869, and there read to the population the Queen's Proclamation transferring the Territories from the Bay Company to Canada, on which date the proclamation would also be published in London and Ottawa. After these instructions had been sent to the lieutenant governor, however, the Canadian cabinet, alarmed by the insurrection, caused the date to be postponed until the following July. Knowing nothing of this, McDougall returned to Fort Garry, hoping that the minority Canadian Party would rise to his support in the name of law and order. They did not. Riel still had the situation in his own control. So nothing happened and Riel remained in power. When, weeks later, news of postponement arrived, McDougall was left dangling in air, a ridiculous figure, proclaiming an authority he did not possess, a governor without principality or province. Determined to suppress the rebellion, McDougall raised a force under Colonel Dennis but, as he lacked any vestige of authority and possessed no more legal status than Riel's Métis or the

Bay's disbanded guards, McDougall was forced to disband his "army" and the impotent lieutenant governor was seriously discredited again. Down to here it had been a burlesque war.

Finally twelve hundred militiamen were sent from eastern Canada to restore law and order. When this force, under Colonel Wolseley, largely composed of angry young Protestants noisily longing for half-breed Catholic blood, appeared before the fort, Riel fled over the Dakota border and the rebellion collapsed as quickly as it had begun, though fifteen years later Riel was to return to what was then the Canadian province of Manitoba and lead a madcap insurrection, for his part in which he was subsequently hanged. Even the return from monopoly to competitive trade, achieved by the entry into Canada of what is now its vast northern empire, could not be accomplished without conflict. Its atmosphere was charged with explosive qualities reminiscent of the bad old days of the fur wars, for recourse to arms has always been the solution to disagreement and the struggle for power on the frontier.

Competition in the trade for furs burgeoned again almost as soon as Canada had moved into the Northwest Territories and the ensuing brawls had been quieted. The reasons for the rapid development that followed did not derive entirely from the new ownership, however, but in at least as great degree from the forward movement of the times. This was the great era of railroad building to the west and, as in the United States, though somewhat later, Canadians set out to link their coasts with bands of steel. Thus the coming of the iron horse to the plains reopened the fur trade between the south and the Mackenzie and its tributary country. When the railways finally thrust feeder lines north, first to Edmonton and then beyond, until the steel finally came down to the banks of the Athabaska, close to Fort McMurray, the change-over was complete and the bulk of the trade of the northwest again began to flow south out of the country to the railheads, thence across the continent to the mercantile centers of the east. Montreal was back on the fur map.

Even darker days were in store for the Old Company. It persisted in sending its freights from the Bay Company overseas to Britain, but mainly these were secured in the eastern arctic. Even the Bay soon discovered that its own easiest route out of and into the Mackenzie Valley was over the water to railhead, thence across the continent to tidewater. The coming of the steamboat finally settled the question and again revolutionized the conduct of the trade in the empire of the Mackenzie. The age of the stern-wheelers had come to the river. Now even the stern-wheelers are moving on to the graveyard of remembered ships.

Came a day around the end of June, depending on when the ice went out of Great Slave Lake, the first of these shallow-draft, paddle-wheeled old ladies would churn past Big Island and enter the Mackenzie River, heading for the Arctic Circle and points north.

Came the end of October the same stern-wheeler, or one of her sisters, would emerge from the big ditch, thresh across the southwestern reaches of the lake, swing through the Slave River delta and clip-clop around its fantastic bends until Fort Smith had been reached.

Like the York boats before them and the canoes of discovery days, they went north laden with goods for trade and came south again carrying mountainous bales of fur.

From 1883, when the first of the thrusters, the *Grahame,* was put together at Fort Smith for the Mackenzie service, the stern-wheelers carried the freight of the Far North and handled its passenger traffic, plying the Dominion's greatest inland waterway system for approximately sixteen weeks in the year, traveling farther than Alexander Mackenzie ever went, passing Aklavik in the delta and steaming on to Port Brabant (Tuktuk) on the outer Arctic coast, a jaunt of 1,700 miles, in which the portage between Fitzgerald and Fort Smith, where the province of Alberta meets the Territories, is the only interruption. Cargoes once moved over a crude tote road on the portage. Today they are transferred in trucks on gravel highways, over which passengers ride in state in taxis. For years the wood-burning *Athabaska* plied the southern end from McMurray to Fitzgerald. The 1,300-mile northern run from Fort Smith at the northern end of the portage to the Beaufort Sea was latterly served by an elderly, waddling beldame called the *Distributor III* and her sister, the *Mackenzie River*, former log-burners converted to oil in their sunset years, with ample supplies available along the line at Norman Wells.

Strange waters these for steamboating, waters which

run through sand-ridden deltas and long river shallows, which carry through more than a hundred miles of Great Slave Lake, and finally lead out of the Mackenzie into the open Arctic Ocean. With the stern-wheelers went loaded barges, pushed ahead, a tricky chore which often necessitated dropping all but one, to shuttle back and forth through stretches of treacherous open water or fast currents and bring along the rest of the convoy. Drawing only three feet of water and as broad-beamed as any Madam Chairman, they lurked in river mouths or in whatever lee they could find waiting for fair weather in which to scoot through Great Slave or out to Brabant. Even so the elements often trapped them. Last time the writer visited the Territories, word came into the gold camp called Yellowknife that *Distributor* was unreported since leaving the sandy jaws of the Slave and must be somewhere on the lake in a wind of gale force. A search plane promptly took to the air. Two hours later the old girl was found, quietly riding it out in a lee made by a jut of land no higher than an adult sandspit. Certainly the ladies have earned the retirement that is coming their way, if it has not come by the time this is read.

Back before oil was tapped at Norman in 1920, or Gilbert LaBine discovered the makings of the atom bomb at Great Bear Lake ten years later, the sahibs of the fur trade rode these ships in solemn majesty, taking their ease under striped awnings aft, where they were lulled by the steady clunk of a wooden-bladed paddle wheel, thrusting them toward the arctic at a terrifying six knots.

The picture has been likened by romantics to Mark Twain's Mississippi. But no. These rivers flow across no lush land, but through a stony-faced country which, rich though it is in fur, oil and minerals, has nothing of ease in its horizons. No rollicking roustabouts ever gathered on its jetties to sing barbershop harmony as they hoisted bales aboard the steamboat. Instead, high-cheekboned, blank-eyed descendants of the Slaveys, Dogribs and Hares poured logs aboard at Fort Providence, Wrigley, or Arctic Red River, or at wayside points which were just woodpiles, to keep steam in hungry boilers. And, for one man's money, the Mackenzie is tougher going than the Mississippi ever dreamed of being. The shallows here are just as tricky. The channels shift every season as glacial boulders in the river bed roll over in their sleep. Then there are those awe-inspiring stretches of open lake and coastal water to be run, often against a clock that says winter will be here any minute.

From 1883, when the *Grahame* first lurched north, until the mineral rushers turned up in the early 1930's, the sternwheelers had the Far North to themselves all summer. Year in and out they pushed their barges into the stream at Fort Smith toward the end of June, themselves loaded to the gunwales with trade goods and traders. Weeks later they returned, carrying huge bales of fur and Bay Company officials, homebound on leave. Downriver, picturesque trappers rode between forts, en route to dicker for the next winter's fur-catching supplies. Missionary priests, government doctors, Indian agents, the occasional red-coated Mountie, sometimes a pretty girl bound down north to wed a Bay man, or a wife returning with her brood to rejoin a husband on some isolated northern post after a winter outside; these made up the passenger lists, these and the fur sahibs and their ladies. Crews were picked up in the country. This was still the fur empire, despite the fact that Canada had moved in and taken over. Nobody who didn't belong down north wanted any part of it – and the fur barons still wanted no part of anybody else in their country.

Then came the first phase of a two-part revolution in an empire where life has been a series of revolutions ever since it was "discovered" in 1789. As always, the new revolution was to change every aspect of living in the Mackenzie Basin, one visible result of which is the retirement of the Old Ladies of the Big River. All this is getting ahead of our story, yet somehow the record of steamboating days belongs all of a piece – and eras always overlap.

The initial augury of things to come was written high in the skies of the late 1920's, when the early air voyageurs of the Far North set out to prove that the Territories were a natural for air travel. First, the renowned Canadian bush pilot, Punch Dickins, flew as far as the marge of Great Slave Lake at Fort Resolution. He was followed in short order by such "name" fliers as Wop May and Leigh Brintnell, the latter, after earlier reconnaissances, carrying LaBine to the scene of his radium strike on Great Bear Lake. Once the fliers had proved the feasibility of their claim, the boats were doomed as passenger carriers to all but the fur barons who owned them.

The mineral rushes of the early 1930's into Great Bear Lake, into Lake Athabaska, and later into Yellowknife and the country northeast of Slave were airborne. But as mines developed, the need for water transport to pack heavy machinery and supplies for the new mines became acute. The old girls could never have handled the job alone. Their Bay Company owners might have expanded the service, adding fleets of tugs and barges, and they might have been able to knock freight rates down to a point at which the bringing in of mining equipment would not have been prohibitive. But, fundamentally, the owners of the stern-wheelers wanted no part of the outlanders who were invading the fur empire in search of minerals. Times might have changed, but not the outlook of the men of fur. True, the newcomers were not seeking to compete for fur, but their coming meant that the fur would retire farther into the wilderness, on the one hand, and that the minemakers would bring a note of modernity into the bush – plus a knowledge of outside prices, dissemination of which to the Indians the fur barons had always shunned like the plague. Shortsightedly, the fur barons held their freight rates prohibitively high, erroneously believing there was nothing the mining folk could do about it. But the miners did something and the stern-wheelers, for the first time, found themselves faced by competition, established by the radium hunters of Great Bear

Lake. That meant that *Distributor*, *Mackenzie River*, and *Athabaska* remained primarily carriers for the fur trade, the big influx of passenger traffic gone to the airplane, their strangle hold on the freighting business of the Mackenzie Basin to the new competitors, gentlemen of highly modern outlook, thinking in terms of oil-burning steel ships, certainly not of voracious old steamboats which devoured every woodpile in the country. The handwriting was on the walls of Fort Smith and Aklavik for every old-fangled tenet of the Far North. That was the first phase.

The second, which administered the coup de grâce, was World War II. It brought, first, the Americans to develop oil at Canol, deep down the Mackenzie across from the wells of Norman, whence they slung a pipeline through the mountains to tidewater, $434,000,000 worth of it! It brought others to build the Alaska highway, while Canada filled the country with outsize airdromes on the staging route to the Orient, cleared landing strips at virtually every minesite and trading post in the Territories. It brought, finally, the opening and intensive development of the uranium mines at Great Bear under Canadian government ownership, which included public ownership of ships in the stern-wheelers' competitor line, a competitor with a bottomless bankroll for new ships, tugs, barges, whatever the job might require. Once the government was in transportation in the Territories on a for-keeps basis, the last hope for return to the good old days of the fur empire went out the porthole. Official Canada had signified its determination to crack open the

Territories and bring out not fur but gold, radium – and uranium for The Bomb.

Throughout the war years the old girls churned silt valiantly. Actually the old *Mackenzie River* had been retired to pasture at the Fort Smith waterside years before. Now she was renovated, chinked up, converted from wood to oil, and put back to work. Throughout the days of the Canol development she and her younger sister, *Distributor III*, flailed the Mackenzie's turbid waters with their butter-churn paddles and poured supplies into Norman with the best of them, hundreds of tons to the voyage, plus all the loaded scows they could shove. That was the last great heyday of the stern-wheelers. They knew they were doomed. They knew that tomorrow on the rivers would be a day of diesel-driven craft, faster and scientifically constructed to cope with the strange assortment of waters and weathers that the long pull down north involves. Yet perhaps it was their finest hour. They ran the lake in storms they would not have dared in the old leisurely days. They carried loads no stevedore would have deemed them capable of bearing. With ice floes crunching against their sides they steamed past Hay River and Big Island at the end of June and into the Mackenzie, drawing every last fraction of an inch of water the shallows ahead would allow. They sneaked home from the season's last voyage as Slave Lake closed in around them, barely making the delta in time for the run up to Smith. If this was to be farewell, it should be a glorious valedictory. It was.

What has followed since World War II is anticlimactic.

The old girls were steaming on borrowed time. The Bay Company, in 1947, was forsaking the transportation business down the Mackenzie, giving up, of all things, to a new monopolist, the government of Canada. Shades of old George Simpson! The Bernhardts of the Big Water were playing their final farewell tour.

Came the end of June, 1947, the old *Distributor* creaked into the stream from her moorings at Fort Smith, loaded with trade goods for distant posts and carrying younger and less austere sahibs of the fur trade on her decks. They said it would be her last year, though she might have to tough it out again in '48, if the government couldn't get new ships into the water in time. Slowly she meandered down the curlicue Slave, which at one point in its journey loops around 16 miles of country to make a net gain of less than a mile in its progress to the lake. When she reached the delta her elderly skipper, who once upon a time took his weather by sniffing the breeze, pondering the skies, and consulting his memory of auguries, studied government meteorological reports before venturing into open water. Throughout the lake crossing and the call at Hay River, the Old Man watched the skies apprehensively, for what their seniors speak of as "guesses" on the part of the young scientists who predict today's weather don't always work out. Ship and skipper sighed together in relief as Big Island was left astern and the old *Distributor* nudged her way into the Mackenzie proper.

Forty-five miles down the line the old girl made her call at Fort Providence, to be greeted by every man, woman and child, white, Indian or what-have-you in the settlement with an enthusiasm you'll never see turned on for the new and shiny steel ships, for the coming of the first steamboat was always the event of the year in the Territories, and something will be missing from the scene – at least a gala day – when the old girls whistle off the jetty for the last time.

Then on she trundled downstream to Simpson, where the Liard comes tumbling in. Wrigley . . . Norman Wells in its gentle grove and the mountains close by in the west . . . across the Circle and on to Fort Good Hope . . . Arctic Red River and soon the big stream breaking up as it enters the delta, down the west arm to Aklavik, that white-painted colony, complete with hospital and tilled fields which could vie with many a New England village in its white tidiness. The outside run to Port Brabant, waited for until the weather set fair, ended the journey. As she started for home on her last trip in September the old girl was probably taking her last look at the Arctic Ocean, the deltas, the big river, all the long miles down and up which she had trudged so faithfully since she made her first voyage down north. The passing of the old stern-wheelers brings another change of life to the Mackenzie country.

Now it's the air age and people can't wait for the *Distributor* or the *Mackenzie River* to come and get them, when they can go from Norman to Fort Smith and on down to Edmonton in a handful of hours. The ways of the industrial age have reached into the Territories, where men are seeking the fastest and cheapest methods of moving their heavy goods over the water during the brief open season. The old girls haven't had a hope since the day LaBine found pitchblende on Great Bear Lake.

Even so, their passing will be mourned. Years hence folks will sit around hot stoves at Resolution and Wrigley, at Simpson and Aklavik, talking of ears that were cocked for the sound of the distant whistle that meant the coming of the steamboat round the bend, and the end of unendurable winter. Oldsters in the company will bemoan all this modern clatter and hurry, while youngsters in the land point out brashly as always that you can't stop progress. Wooden steamboats, they'll say, and there'll be scorn in their voices. Whoever heard of wooden paddle-wheelers! Why, that was in Mark Twain's day, down on the Mississippi!

But, like the canoe and the York boat, they played stellar roles in their day. They carried the goods of the country and they carried its people when no other alternative offered. They did it faithfully and well, and when the chips were down and we needed pipelines and airfields and machinery to get the uranium out, the old girls came out of retirement and put their backs under the burden! But now to the tale of the airplane and the tractor train, the highway and the airdromes, staging posts on the way through the new Northwest Passage.

Sic transit gloria . . .

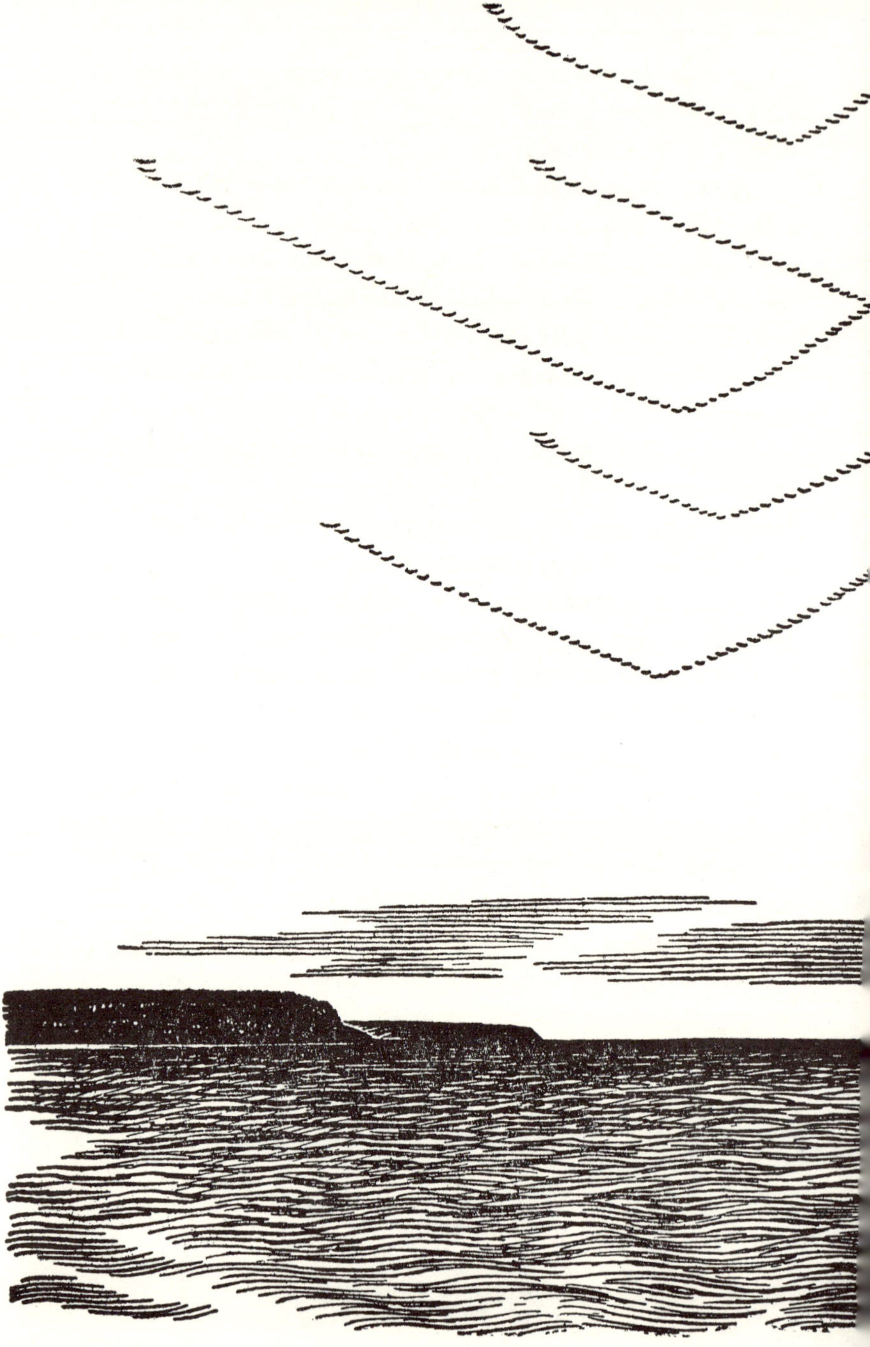

CHAPTER EIGHT

The New Empire Builders

Brintnell came into the bedroom and said he felt sorry about the weather. Outside, the morning atmosphere contained all the ingredients of soupe aux pois, with the possible exception of chopped carrot, the resultant witch's brew having been tipped into the treetops, so that it was possible to see almost across the street from a second-story window. Inside, four citizens in flannel shirts, breeches, knee boots and windbreakers gazed at each other sullenly, for this was the third morning on which they had risen with the dawn, hoping to fly to a landing on distant Lake Athabaska.

"What are the chances, Leigh?" one of them wanted to know. "Liable to be able to fly later?"

The pilot said he hadn't the vestige of an idea. There'd be another weather report from the north in an hour, but all the omens were sour. Pea-soup fog at Goldfields; a forty-miler and heavy snow at Chipewyan; gale and blizzard at Fort Smith and Resolution; visibility and ceiling zero everywhere, with no sign of a break . . . Definitely not good. So after breakfast the quartet walked down to the airbase, as they had walked yesterday and the day before yesterday, to gaze dolefully at eleven idle planes standing on the ice of the Clearwater. Thereafter they patrolled the town again, calling at the government telegraph office, the drugstore, the Hudson's Bay post, the hotel, the airbase again, and so on and on ad nauseam. And the morning and the evening were the third day of weatherbound incarceration in Fort McMurray.

The fourth morning dawned crisp and clear, however. From Rae and Resolution on Great Slave Lake the radio brought news that the storm was breaking and the ceiling lifting. The equinoctial gale had blown itself out. Six o'clock found us wolfing eggs and bacon in the Franklin Hotel. At six-thirty, packsacks, sleeping robes, and all the assorted dunnage of the northern wayfarer were weighed and piled beside the cabin of an air freighter. By the time the seven-o'clock weather report gave "All clear," the travelers wedged in among an assortment of packing cases and personal effects, ready to take off for a new settlement where seekers after gold had changed barren caribou pasture within the space of a year into a habitable industrial community.

Goldfields was its name. It lay nearly six hundred miles

north and east from Edmonton, and approximately half that distance northeast from the steel's end at Fort McMurray. It was then little more than a year old, but on the scene of its principal activities you could find electric light, running water, and steam heat; while among its urban qualifications were listed a hotel, two radio stations, stores, a post office, a doctor and a schoolma'am, church services, bridge-table covers, and dog-team taxicabs. A booming mining community – and just up the shore, a few minutes' flight away, the dingy semisqualor of the old fur traders' headquarters, Chipewyan, unchanged by time.

Goldfields had happened because an elderly Scandinavian trapper liked to go prowling during the long afternoons of the off season. For years Gus Nyman had dreamed of gold and had acquired the habit of pottering about among the rocks adjacent to his summer habitat, in the hope that someday he might find the open-sesame to wealth. From one such expedition he returned bearing a number of pieces of reddish rock liberally shot with stringers of quartz. Nyman had no idea what it might be, but he possessed an unconscionable curiosity to find out.

Subsequently his path crossed that of an old-time prospector, to whom he showed his samples. The prospector pricked up his ears, listened, and took out his miner's glass. Later the duo visited the scene together, chipped off more samples. The prospector suggested that he should proceed outside, procure assays, and if gold values of moment should be returned, endeavor to interest capital and come back to stake the ground.

Nyman agreed. The prospector went away. Time passed. Nyman began to think there had been nothing to his dreams after all. Conjecture became certainty when one morning he awakened to discover that a staking party had arrived and that his ground had been claimed for one of the greatest mining corporations on the face of the green footstool. The profit to Nyman? Exactly zero, unless you count the naming of the bight of water adjacent to the strike Nyman Bay something to mark down on the profit side of the ledger. It wasn't the company's fault. They had never heard of Gus, and no stakes marked his claim to fortune. Nevertheless, he had laid the fuse that fired the charge to set a gold stampede in motion. That is how Goldfields was born.

My own first view of it came months later, after flight from Edmonton and McMurray down the Athabaska River en route to the radium strike beyond the Arctic Circle at Great Bear Lake. That was in March, 1936 – and that contrast between new and old, Goldfields and Chipewyan, remains unforgettable. It is the key to what has happened down north. That it was a fur trapper who found it, didn't know what it was, or what to do about it, and so got nothing out of it, somehow was typical too.

Visualize now the shores of Great Bear Lake in 1936. Where Gilbert LaBine discovered radium-bearing rock in 1930, a modern mining plant was in constant operation.

Shafts had been thrust down to depth. A concentrating plant capable of handling a hundred tons of rock a day had been installed and was running full blast. A hundred men were permanently employed. The bunkhouses were sanitary, comfortable, modern dwellings. Billiard tables, a library, and a tobacco shop were included in the equipment of the recreation hall. A company-owned flying boxcar traveled with timetable precision between the camp and the end of steel, hauling radium concentrates on southbound journeys and supplies – including fresh meat and vegetables – on its northward trips. Three thousand miles away, at Port Hope, Ontario, a radium refinery had been established, where concentrates were reduced again to the marketable form. Gilbert LaBine's wildcat of 1930 had become not merely a producing mine but one which supported blue-ink entries on the ledgers in the company office.

But this is not all. Not far from the Eldorado workings was a town where all the amenities of everyday life could be found. Scattered about the vast outline of the lake, prospecting and developing parties were at work. As in other camps in these amazing Northwest Territories, telegrams sent in the morning from the Dominion government's wireless station often brought answers before the day's business was done. If the weather was propitious, it was no trick to breakfast at LaBine's mine, lunch at Fort Smith and dine in Edmonton, a thousand miles to the south. Schools, churches, trading posts, doctors, daily market reports, bridge parties, cooking to compare with the cuisine of the effete East, air mail – all these and many more comforts and services were to be found beyond the arctic's rim in Great Bear Lake in 1936 – and around the point in another arm of the bay, as everywhere in the country, it seemed, the inevitable, half-ramshackle trading post, constant reminder of past glories.

The story of the airplane and the miner in the Mackenzie Northwest are one and indivisible. Without the bush pilot and his pontoon- or ski-shod aircraft it is eminently doubtful if the mineral resources of the Territories would have been cracked open to this day. Without the miner and the traffic he has brought to the hinterland airlines, it is equally doubtful if the bush pilot could have stayed with the country. The first pilots who flew this wilderness trail, Dickins, May, Brintnell, Gilbert, Berry and others, were the modern counterparts of such men as Hearne, Dease, and Dr. John Rae. Men of the stamp of Gilbert LaBine had the touch of Alexander Mackenzie himself. Others had the flair of Governor Simpson for administration and shared his horror of entries made in ledgers with red ink.

But it was the fliers who began it. Without them there would be no golden north to this day. If a Punch Dickins hadn't been convinced that the proper way to deliver the mail in the Mackenzie country was by air, not with dog sleds in winter and paddle-wheel steamboats in summer, it is reasonable to assume that LaBine would never have reached Great Bear Lake to find pitchblende, the mother rock of radium, uranium, and U-235. If there had been no airplanes, no Dickins, no May, no Brintnell, and this one man LaBine, then there might have been no atom and no Hiroshima. But that is a story which comes later.

The beginnings of aviation in the Mackenzie country are dated 1921, when developers of the oil wells at Fort Norman decided to experiment with the possibility of flying supplies and men to their subarctic operation. Two Junkers low-wing monoplanes were purchased in the United States and flown to Canada. The town of Peace River, on the water of that name, was chosen as a base and a route laid out which followed the Peace north to Fort Vermilion, then cut across country to reach Great Slave Lake at Hay River, whence the Mackenzie was followed down to the oil-well site.

The planes set out from the Peace on March 24 and reached Fort Simpson, where they were grounded by engine trouble and ski and propeller breaks. A month passed before the Junkers were fit to fly again, by which time impending breakup of winter farther south had become a problem. If they reached Norman, would they be able to get out again and back to the Peace while solid ice held in the river for a landing? The two pilots, Elmer Fullerton and George Gorman, decided not to run the risk. They returned to base from Simpson and the experiment was abandoned. The beginning had not been auspicious.

Five years passed. In 1926, H. A. ("Doc") Oaks, a pioneer bush pilot who had been attempting to establish commercial service into the Red Lake mining field in northwestern Ontario, went looking for capital and in the course of a visit to Winnipeg called on the late James A. Richardson, millionaire grain man and industrialist, and invited him to become an aviation magnate as well. It was just the kind of notion to appeal to a man of Richardson's adventurous temperament. Oaks walked out with a check, and Western Canada Airways had come into being. It was Western Canada that cracked open the door to the Mackenzie empire and wrote into the glowing record of the air the names of Dickins, May, Gilbert, Brintnell, and all the others who were its pioneers and the corps d'élite of the Canadian bush pilots.

Young Dickins was a man of ideas and, as time wore on, became a young man of one consuming idea. That idea, simply stated, was that the isolation of the fur empire could be broken wide open by air. The problem was to convey his own enthusiasm to people with business to do in the country.

It was January, 1929, before the young enthusiast made his point. On friendly terms with postal officials in Edmonton, he learned that they had on hand a pile of delayed mail, much of it Christmas parcels for fur men in the Territories, which it was far beyond the capacity of the overland delivery service into the country to handle. If no way could be found to move it, the people of the Mackenzie would not get their 1928 Christmas parcels until well into the summer of 1929. Dickins offered to fly it in. The postal authorities in Edmonton queried Ottawa, which authorized an experimental flight as far as Resolution. The cabin of Dickins's plane was loaded with mailbags, a postal official climbed aboard to act as postman, and the first air mail north was on its way.

Weather held them at Fort McMurray. In those days

when bad weather came down out of the north, you waited it out and pushed on when you could, without knowing what lay ahead. If you ran into a blizzard, you cut back to base if fuel supply was sufficient to permit the return flight; if it wasn't, you sat down on the first lake you could get into and waited. On this by-guess-and-by-God procedure, ten days were consumed in getting the mail as far as Fort Smith. Resolution was made the next day and the last sack was unloaded. Then Dickins took off light, pointing for the Mackenzie and Fort Simpson, to tackle another experiment – the southbound freighting of fur.

The party reached Providence southbound with the thermometer registering 62° below. They pushed on back to Resolution, where in landing on the rough ice off the settlement they struck a hummock which damaged their undercarriage and cracked one blade of the propeller. The great experiment had collapsed. Dickins, the zealot, saw tragedy ahead. This is just what the Jeremiahs had said would happen. If the first flight north ended with a cracked-up airplane left behind, and the crew making its way ignominiously back to railhead by dog team, or the first boat, the we-told-you-so brigade would be in full cry against what it had already branded a harebrained scheme and aviation would get a setback from which it might not recover for years.

Dickins and his flight engineer, Lew Parmenter, looked over the wreck and wondered what to do next. Ironically, aid came, of all sources, from the blacksmith shop of the trading post, from an employee of people who wanted no part of airplanes flying around their back-country empire. The blacksmith said he could fix the broken undercarriage with riveted water pipe. That left the propeller. Could they cut it down and get its blades back in balance? Pilot and engineer decided to try. For days they filed and tested, listened to the rough roar of an unbalanced prop, filed again and tried it again. Finally they produced the sweet hum of blades in balance and were ready to go. They reached McMurray without further mishap, where they picked up a new propeller, shipped north by the weekly train.

Dickins stayed in the north all that winter, based at McMurray. In March he picked another load of mail off the train at the end of steel and delivered it down the line as far as Fort Good Hope. Unloaded, he detoured before heading south and became the first North American flier to cross the Arctic Circle. When he came out the pilot recommended making Norman the northern terminus for winter flying. North of the wells, he said, with only three hours a day of half-daylight, the business was too risky. A few days later he was landing at Fort Rae, at the tip of the northern arm of Great Slave Lake. Before breakup he had made two flights into the Yellowknife section near-by, later to become the scene of one of the greatest gold rushes in the history of the north.

That summer Dickins made the first flight down to the delta, almost exactly one hundred forty years after Alexander Mackenzie's voyageurs paddled him through its waters.

The new discoverer landed at Aklavik, to be greeted by Eskimos whose proverbial stoicism was shattered by sight of the great bird landing on the water and disgorging human beings. The Eskimos immediately began to ask Dickins to take them fiying in the "great bird" and for four hours following his landing the young pilot took load after load of natives aloft in a series of ten-minute flips, despite the fact that for each of the past five days he had been at the controls for more than nine hours.

During this same summer another pioneer flight of note was made into the Territories by one of the airmen whose names will always be associated with the opening of the arctic's resources, Leigh Brintnell, the pilot of record in the opening paragraphs of this chapter. What follows is culled from the official log of a memorable pioneer expedition, as laconic in its description as was Mackenzie's journal of his own voyage down the great river. Says the official record:

> On August 5th, 1929, W. L. Brintnell left Winnipeg on a tour of inspection and reconnaissance covering the Mackenzie River area, the Yukon and the littoral of British Columbia.
>
> The first stage of the flight was to Cranberry Portage – a distance of 450 miles – into the heart of Northern Manitoba, thence across Saskatchewan and, on the same day, to Fort McMurray on the Athabaska River, making 920 miles in nine hours flying time.
>
> On August 6th the flight was continued to Fitz-

gerald and to Simpson on the Mackenzie River, a distance of 750 miles.

On the 7th the machine was flown to Norman and in to Great Bear Lake.

From the 8th until the 16th fifteen flights were made for the purpose of reconnaissance over great Bear Lake [and] in the Norman area. 2,480 miles were covered during this period and visits were paid to Boland's Camp, Long Point, Dease Bay, Dease River, east of Great Bear Lake, and McTavish Bay.

On August 17th Norman was left for Good Hope, Arctic Red River and Aklavik, a distance of 435 miles.

On the 24th the machine set forth on the second main stage of the flight from Aklavik, south along the Husky River, west along the Rat River, over MacDougall Pass, 1,150 feet, and west along the Bell River; still west along the Porcupine River into Alaska, to Fort Yukon. South along the Yukon River, back into the Yukon Territory, and on to Dawson, a course of 742 miles from Aklavik. This was the first flight from Aklavik to Dawson, the time taken being six hours and thirty minutes.

Dawson was left early on August 25th. The flight was continued south, following the course of the Yukon and Lewes rivers, via Carmacks, Little Salmon, Big Salmon, along Lake Laberge and White Horse to Carcross, a distance of 450 miles.

On August 26th, Carcross via White Pass, Skagway, Juneau and along the inland water channels to Prince Rupert, a distance of 500 miles.

From Prince Rupert on August 27th inland, following the line of the Canadian National Railways to

> Prince George and on east to Edmonton, a distance of 980 air miles.
>
> On August 28th Edmonton was left for Emma Lake, north of Prince Albert, and on the 29th onward east to Cranberry and thence to Winnipeg, 1,225 miles.
>
> The flight occupied 95 hours of flying time, during which 9,000 miles were flown, a distance equal to a flight from Ireland over the Atlantic Ocean, across Canada and over the Pacific Ocean to Japan. Fog was encountered on the White River, low clouds and rain at Carcross. Clouds and fog were also encountered at White Pass. With the exception of haze and smoke from forest fires at the end of the journey, no serious mishaps were experienced, although 1,000 miles were flown inside the Arctic Circle.

The new empire builders were beginning to fill in their maps, precisely as Mackenzie, Franklin, Back, Dease, Simpson, and Rae had etched in the blind spots almost a century and a half earlier. But the newcomers could fill more in a season than their forerunners in the country could do in a decade on the ground, learning meanwhile more of the country's topography than their predecessors had ever learned.

Northerners themselves were the first to welcome the breaking down of barriers, almost impenetrable during the long winter night. In the records of Western Canada Airways (later to be merged into Richardson's continent-wide Canadian Airways) is a letter from a government Indian agent commenting on the first mail flights and the boons that aviation had begun to confer on the country.

"I received more mail in January [of 1929]," he wrote, "than in the last four winters combined. Every winter I have to visit, only once, due to expenses beyond reason, Providence and Hay River, a 150 mile trip with dogs at a cost of $400, with misery during the twelve days the journey takes. This year I made it at a cost of $135, for the round trip, including excess baggage, which means I can visit my district three times in the next winter for the price of one former visit and enjoy a pleasant journey instead of hardship.

"To go out in the winter used to cost a fortune. Last winter I went from Resolution to Fort McMurray at a cost of $400 with misery. This winter D. K. Murdoff of the Northern Traders Company went in three hours with comfort, caught his train, while I missed mine by three hours and waited a week, after spending nineteen days on the road."

Such correspondence clearly indicates what the first flights meant to the people marooned in the Far North. A journey of nineteen days on the ice, riding a dog-drawn sled, reduced to a three-hour flight in the cabin of a heated airplane! Mackenzie's hundred days on the great waterway cut to a two-day journey from Chipewyan to Aklavik and back. On all this the barons of the Bay Company looked askance, as they did on any activity likely to bring the amenities of settlement into their empire, for as people arrived the fur would leave for open spaces farther on. The smaller, independent traders, however, were of different mind. Here was a form of transport that would enable them to get their furs to market and, by slashing the time between purchase and sale, permit them to turn capital over much oftener.

Now the stage was set for new activities that were to bring a new breed of men into the country in search of natural resources and mineral wealth. Thanks to them and the airplanes that brought them, mines have been developed and modern communities opened, as far as a thousand miles north of the northernmost transcontinental railway.

The rush began because Gilbert LaBine walked into Jim Richardson's Winnipeg office in 1929 and sought advice on an aerial prospecting expedition down north. In those days LaBine and his associates held a charter for a company known as Eldorado, which had been exploring a group of claims in northern Manitoba. The property had failed to keep its original promise and, while he still had money in the bank, LaBine proposed to go out and find something to take its place. Could one of Jim's boys fly him into the Bear Lake country, leave him on the ground and be sure to come back and get him a few weeks later? Richardson said it could be arranged.

The arrangements took LaBine north on the first leg of the wide swing around the Mackenzie Basin and over the Rockies to the Pacific Coast, already described in the log of Brintnell's journey. The detour to Great Bear Lake had been made, in fact, for the precise purpose of dropping the prospector on its shores, Brintnell carrying on, back to the Mackenzie and on to Aklavik, leaving the mineral hunter on the shores of the great inland sea, to be picked up weeks later by Dickins, who had undertaken to come in from McMurray to fetch him.

LaBine was not the first prospector in the region, however. Ten days earlier Dickins had accepted a charter from two Americans who had stepped off the train at the McMurray railhead, a man named Moore from San Francisco and his partner, McIvor, of Butte, Montana. Thus Dickins became, by a margin of days, the first man to fly in to Great Bear. They went in by way of Norman, where high winds delayed their departure under conditions in which Dickins knew it would be fatal to attempt a landing on the huge body of open water. As the next step the pilot made the 80-mile hop across to Boland's Camp, where the lake empties into the Great Bear River and heads downhill toward the Mackenzie. There they were grounded for several hours by rain, but finally chanced the overwater hop to Dease Arm, in the northeast corner of the lake, almost two hundred miles away, where the two Americans proposed to make the site of old Fort Confidence the base for a prospecting expedition.

Sleet caught them in mid-lake and Dickins turned to get back to Boland's with all haste, but the storm closed in, icing his wings and windshield and forcing him down into the nearest semisheltered water he could find, a deep inlet halfway up the west side of the lake, called Deerpass Bay. The pilot came in safely and the seaplane was tied down while the party waited out the storm.

Next morning the weather had cleared, though a heavy swell was running, as prospectors and pilot taxied down the bay to swing into the wind and tear into a take-off. Just as Dickins gave the engine the gun, however, a heavy swell hit the pontoons, a front strut buckled, and the party was immobilized, barely making shore as the passengers fended off rocks while Parmenter wedged a pole between wing and float to hold the ship in one piece. Then they paddled into a sandy refuge in the lee of high ground, where Dickins and Parmenter, by this time accustomed to improvisation, after their experience at Resolution during the previous January, fashioned a new strut from poles of spruce. Leaving their passengers on Deerpass Bay, with a promise to come back or send somebody to fetch them, Dickins and Parmenter limped across country to Simpson on a wing and a prayer to do a blacksmithing job that would hold their aircraft in one piece until they could get to McMurray and wire home for a new strut. It was Brintnell, northbound with LaBine, who brought it and winged on, promising to pick up Dickins's Americans at the same time he dropped LaBine on Great Bear.

Brintnell took LaBine in to Great Bear and dropped him at Boland's Camp, then flipped across to Deerpass Bay to pick up the two Americans, who by that time had seen all they wanted of the rugged country in which their pilot had left them. Brintnell took them across to Simpson, then returned to the big lake to pick up LaBine and a government geologist, Dr. Bannerman, and ferry them to favorable prospecting localities around the shore. During the next few days the duo carried out an aerial reconnaissance of more than a thousand miles around the coast of the third largest body of water in North America, Brintnell finally leaving them on McTavish Arm and heading off for the Mackenzie, Aklavik, and the long swing around the Yukon and the British Columbia coast.

LaBine and Bannerman prospected the surrounding country for three weeks. On August 26 Dickins flew in to the prearranged rendezvous and picked up the two men. As they flew out, sweeping low over the country, intent on the rocks beneath their pontoons, LaBine saw beneath the wings what he had been looking for ever since he came into the country, the strangely stained rock known as cobalt bloom, in which later he would find pitchblende, mother of radium, uranium, and U-235. The scientist-prospector sat silent in his seat as the plane roared south. But he'd be back in Bear Lake before the ice went out in the spring. If his mind had not been made up before, it was now.

Such was the record of the first season of air travel in the Far North. Already, as in eastern Canada, the airplane had become the transport of the seekers after mineral wealth. The mining men had been waiting for this. Geological reports on file in Ottawa clearly showed the region to be heavily mineralized, an area as favorable to prospecting as any in the country. But down to here it had been too inaccessible, a three months' journey for a man traveling light; a country in which water transport routes swung away from the mineralized regions when the steamboats turned west at the mouth of the Slave; a country into which it would be almost impossible to move machinery and men, at least until improved transportation could be found. But if you could use the air to move men and their basic equipment, if you could make prospecting parties mobile and keep them fed by flying in food, then you might turn up something that would justify the expenditure on improving ground-level transport. Which is precisely what happened.

On March 28, 1930, Dickins flew Gilbert LaBine and his partner, E. C. St. Paul, back in to Great Bear Lake by way of Great Slave Lake and Fort Rae at its northern tip. When they said good-bye on the McTavish Arm, a date was made for a September rendezvous at the same place. In waist-deep snow the two men hauled their supplies ashore on the sled they had brought with them, pitched their tent across a narrow gut of water from what is now called LaBine Point, site of the most remarkable mine in North America. And it was while they were there that the miracle happened.

LaBine and St. Paul prospected the shores of McTavish Arm as the days grew longer and the strengthening sun turned the snow-clad lake into a solid glare of brilliance. Down to then they had found only signs of interesting mineralization on the higher ground, where the sun had cleared the surface of the rock, but nothing that could be called more than a sign. Then, one April day, as they were mushing home along the edge of the lake, the sun-glare caught St. Paul in the eyes and he went snowblind. In agonizing pain the prospector stumbled the rest of the way to the camp, where he lay in pain for days, his eyes covered with the bushman's specific, poultices of wet tea leaves.

Meanwhile, between poulticings, LaBine moved around the immediate area, never without his short-handled prospector's pick, chipping off samples of rock, screwing a magnifying glass into an eye and peering knowledgeably at what was to be seen in them. A week after St. Paul had been laid low by the blindness and was approaching convalescence, LaBine came back to the tent to make a poultice change, carrying samples of a dark-gray rock in which occurred small bubblelike blobs of black, each with a peculiar dull sheen, which he had chipped off boulders out toward the extremity of the point. LaBine had found pitchblende.

Probably not two prospectors in the whole of Canada would have known the stuff if they had seen it. But Gilbert LaBine was something more than just another prospector. The geologists would admit that on the practical side he is a more knowledgeable man than a good many of the college-trained professionals. Starting as a prospector he had spent most of his youth in and out of the bush. Bitten with the idea that if a man wants to go anywhere in mining he needs knowledge of more than the ways of the wilderness, he had educated himself in the tasks of mine-making and had become what might be called a combination of prospector and executive, president of his own company as well. Later, when he could find leisure, being of an incurably inquiring mind in the ways of rocks, he had taken himself to Queen's University to attend classes under a famed geology teacher, Dr. W. G. Miller, where he learned about pitchblende and other rare rocks.

Most of LaBine's classmates hadn't been interested in the rarer phenomena. They wanted to know about gold and copper, lead and zinc, and the association of these with other rocks, so that a man might know where he was likely to find them. But pitchblende – who wanted to know about pitchblende, when the nearest deposit of any importance was at a place called Joachimstahl, away over in Czechoslovakia? But LaBine wanted to know and he asked questions. Could the doctor give him some idea of the circumstances in which pitchblende might occur, of the rock associations in which a man might come upon it? Was there any region in Canada that the doctor regarded as favorable to its discovery? It seemed there was. Up around Great Bear Lake, perhaps, where the general geology was sympathetic.

In Ottawa, LaBine read reports by other famed geologists, Drs. Camsell and Bell, and from his reading and what Miller had told him he believed firmly that the 'blende might be found along the Circle on the marge of the huge lake, if a man only looked in the right place. To say that he had come there solely to look for pitchblende would not be correct. LaBine was looking for minerals, for anything a mining man would call commercial: copper, silver, gold, whatever the hills might hold. But it was typical of the man, after closing down his northern Manitoba failure, that he should turn to Great Bear, realizing that something new was on the verge of happening, the extension of the air routes into the far northwest. It was the inquiring mind, the urge to find something new in a new corner of the wilderness, that had sent LaBine to Jim Richardson, as it was LaBine's discovery that touched off the first wild mineral rush into the Territories and loaded Richardson's aircraft with a new kind of voyageur and his gear.

It would be correct to say that LaBine had been looking for pitchblende, but perhaps not that he expected to find it. But find it he did, thanks to the sun that struck St. Paul snowblind and turned up North America's most amazing mine, a hundred yards across the water from their own camp.

To LaBine the time from April to September, when Dickins, or one of his fellow pilots, would be coming to get them, must have seemed like eternity. But there was plenty to do. First they must stake as much ground as their two mining licenses would permit. Then they must prospect as great an area as possible, in order to know where to add to their staking the following season, when they would bring a larger party in. After that they must load sacks with samples, checking each chunk of rock, so that they might know where they had taken it. After that they'd be ready to go out to the railroad tracks and east to Ottawa in a hurry, where they could have their samples quietly checked in a government laboratory. The summer may have been eternity to a man who had made one of the important discoveries not merely in Canadian but in world mining history and who longed to get back to civilization, where he could do something about it. But by the same token it went by on wings in LaBine's urgent haste to acquire as much information as possible about the strike before leaving it for the winter. When they moved, it must be with the deepest secrecy. The mining world must know nothing about this until they had blanketed the country around their original claims with as many more as they could legally register.

It was not Dickins who came to pick them up in September, but another famed northern figure, Walter Gilbert, then spending his first season in the Territories and, at the time of arrival at Great Bear for his passengers, on the way home from a memorable and history-making flight. While Gilbert LaBine had been dealing in miracles on the shore of Great Bear Lake, this other northern adventurer had been writing history of another sort by making the first flight over the North Magnetic Pole, a feat descending in straight line from the epic adventures of Franklin and his brothers in the polar wastes a century and more earlier.

On July 1, 1930, in his first flight of his first season in the Territories, Gilbert had flown a pair of salesmen north beyond Aklavik and out to Herschel Island in the Beaufort Sea, where they hoped to persuade the Eskimo population to order supplies of tobacco and clothing for delivery the next season. Whether they were or were not successful in this strange venture is not noted in the record. All that is noted is that Gilbert flew them back to the railhead at McMurray and turned his prop back into the north.

Some time before, government officials in Ottawa had written Brintnell, as manager of Western Canada Airways, asking a price for a charter flight north to the Arctic coast and out to King William Island. Brintnell quoted a figure of $5,335 for a journey which would have required a year of arduous travel on the ground, if it could have been done in that time, to be compressed into approximately a month of intermittent flying. In June, 1930, the government authorized the expenditure and sent out Major Lachlin Burwash, mining engineer and noted arctic traveler, to survey the country from Great Bear Lake north to the coast, thence to carry on to King William Island and seek new evidence relating to the fate of Franklin and his party, based on a report that a large number of that party, not previously accounted for, had perished near Matty Island, close to where it had once been believed the Magnetic Pole was situated. It was also believed that Franklin's own tomb might be found on King William Island.

Burwash traveled as far as Fort Norman by boat, where Gilbert picked him up. In Gilbert's party, in addition to his own flight engineer, Knight, were another pilot and engineer, Buchanan and Gilmour, taken along as spare crew for a plane abandoned by an aerial mineral exploration party far over east the previous season. If the plane could be found and it was flyable, Buchanan and Gilmour would bring it out. It was one of those long shots which sometimes come off, but more often don't.

After an interval on Great Bear Lake, during which

Burwash scouted the country, the party took off for the mouth of the Coppermine, where Burwash decided to remain for several days to look over the immediate area on foot. The two pilots and engineers took off to search for the abandoned aircraft, stranded somewhere along the shore, and before the day was out, they had found her, drawn high and dry on the beach where her crew had left her the year before.

Within two and a half hours, the four airmen had loosened the engine, turned it over, revved it, and Buchanan was taxiing out to sea for take-off. That night two planes were tied down at the mouth of the Coppermine. Next day they groomed the recovered ship for the long journey home and Gilbert's for the proposed flight to King William Island and the Magnetic Pole. Buchanan took off. No sooner had he left than Gilbert ran into cylinder trouble, and decided to beach his plane at the Coppermine for the winter and ship all government gear out on the Hudson's Bay supply ship *Baychimo*, then riding at anchor off the river mouth. Next day *Baychimo* left and Gilbert was preparing for a long winter with the Eskimos when out of the sky appeared the other plane (returned for reasons unexplained by the record and probably forgotten by those who participated in the incident). An insurance policy would call it an act of God.

Now Gilbert switched planes and took off over the sea to the supply ship's next port of call, to recover the government equipment. Bad weather intervened and contact was not made. Meanwhile Buchanan and his flight engineer had recommissioned Gilbert's plane and by the time he was ready to continue north with Burwash in the salvaged ship the second pilot was able to fly the cripple south.

Gilbert, Knight, and Burwash left the Coppermine early on September 4, Knight at the camera as they headed east to Cape Krusenstern, then across the straits to Lady Franklin Point on Victoria Island and along the coast to Cambridge Bay. The next day they swept across Queen Maud Gulf, made a landfall at Cape Crozier on King William Island, and landed in the bay that Amundsen had christened Gjoa Haven, for the ship in which he made the water crossing of the Northwest Passage in 1905. From Gjoa they made the first flight over the Magnetic Pole, until then believed to be a moving point in a 30-mile radius, the pinpoint location of which had been disputed by Canadian and American scientists.

Although they were in and out of fog most of the day, Knight was nevertheless able to compile a complete camera record of the coast of Rae Strait to Cape Hardy on Matty Island, across to Boothia and on to Cape Adelaide, the first photographic check of any section of the country north of the Coppermine, if not of Great Bear Lake.

Almost as soon as Gilbert had completed his run through the Magnetic Polar zone, fog descended sharply, as it can close in in a moment anywhere in these strange waters, which the pilot describes in his log for the day as "the last created portion of the world, necessarily left incompleted through shortage of time and materials."

Sharply they turned southwest, making for the mainland, crossing icepacked Sir James Ross Strait and following the north coast of King William Island to Back Bay, over a country and a sea in which every land and water name is a reminder of the early explorers who came this way.

At Gjoa Haven they had been joined by a motion-picture cameraman, who had come by sea, and the party set out to check new theories concerning the fate of Franklin and members of his ill-fated party. They examined the Franklin cairns, found years before, and explored terrain in which it had been thought possible that new information might be found, but if it held secrets, the arctic would yield none of them. As they left for the mainland, the Franklin record was as it had been before they came.

What the Gilbert-Burwash expedition had done, however, was to provide valuable addenda to maps first written by such heroic figures of the north as Franklin, Dease, Simpson, and others of the gallant little band that had forced its way into the polar barrens early in the nineteenth century. This was a camera check of the coast from the Coppermine far to the east and the shores of King William Island, an invaluable addition to man's knowledge of the land beyond the rim of the continent.

His photographic chores finished, and with that apparent casualness which conceals the urgent sense of responsibility that is the hallmark of the arctic pilot, Gilbert set out for Great Bear Lake to pick up passengers who, while the fliers had mapped the Far North and sought new clues to the fate of Franklin, had themselves discovered and staked one of the most valuable mineral occurrences in the world – the pitchblende of Great Bear Lake. The quintet who hauled sleeping bags, packsacks and snowshoes out of an airplane cabin at Fort McMurray several days later might truthfully be described as the most important planeload ever to come out of the north, in terms of their achievements of that brief summer.

Dickins's experimental runs north at the beginning of 1929 had led to the granting of the first air-mail contract for the Mackenzie country by the Canadian government. Strangely, however, it did not go to Dickins's employers, Western Canada, but to a new company, Commercial Airways, organized the previous year but short-lived. Commercial rendered one great service to the north in that it brought into the region a group of pilots who stayed with the country after Jim Richardson took the bankrupt firm out of pawn in 1931 and merged it with Western Canada and other groups in his dominion-wide Canadian Airways organization.

Outstanding among them and, when this is written, one of the principal executive figures in the postwar aviation world of the northwest, was Captain W. R. May, known to cabinet ministers, mining giants, financiers, trappers, and roustabouts by no name but Wop. Wop May brought something besides flying ability into the north. He carried in his personality what can only be described as a flair for the picturesque and the dramatic. Where Wop went things happened. When an outpost settlement sent out an urgent call for diphtheria serum it would be May who turned up, not in conventional, ski-shod cabin plane, but in an open-cockpit job on wheels, to make a landing in deep snow in subzero weather!

Drama stalked the man. When Commercial took the air-mail contract, its operators decided that the Far North would have its Christmas mail by Christmas for the first time on record. With this lofty end in view, two planes set out together from McMurray on December 1, with May at the controls of the senior ship. But mishap followed mishap. If the weather was fine, one of the aircraft would choose that day to be temperamental. When both planes were in rugged health, the weatherman cracked down. Three weeks after leaving the end of steel the expedition had only reached Fort Simpson, where, with only two days to go, May and his fellow pilot decided to make the run to Aklavik that day, come hell or snowdrifts. They made it to Norman and dumped their sacks. A couple of hours later Fort Good Hope had its Christmas packages. Then the going turned too rough to be by-passed and the planes pulled into Arctic Red River on Christmas Eve, to find the Hudson's Bay post burned to the ground and the trader dead.

Groundbound, the crews of the two planes dug out their Christmas turkey and decided to make the best of their frigid circumstances. The bird had frozen solid, far back in the cabin, and not enough fire could be found in the tragic little settlement, only yesterday ravaged by flames, to thaw out the Yuletide feast. May's party reached Aklavik on Boxing Day. The Christmas mail was late again.

May's greatest exploit, from the point of view of a public which sees the Far North through the dilated eyes of Hollywood, and one which spread his name in continent-wide headlines for a fortnight, was made to order for a sensation-loving press and its sensation-hungry readers. Before it was a day old, headline writers had dubbed it the Mad Trapper Story and it was what can only be called a "natural" for feature writers who knew nothing of the setting against which the drama was played. Fifteen years later, the crime writers were exhuming it to garble it into the new versions of printed drivel, whereas the facts, told straight and in sequence, make it in its own right one of the great tales of violent crime and punishment of the period.

The Mad Trapper's name was Albert Johnson and he was neither trapper nor mad. Johnson lived in a cabin on the Rat River, 70 miles out of Arctic Red, a solitary man and peculiar, but no madman by any yardstick. Johnson was a petty scoundrel, addicted to interfering with the traplines of honest men, an activity comparable to cattle rustling in the old West. His victims finally reported him to the Mounties, who are the law north of the railroad tracks, and the Mounties set out up the Rat River to investigate. That was late in the fall of 1931. When the police reached the cabin, they could find nobody about, so they returned to Aklavik, armed themselves with a search warrant and arrived back at Johnson's shack on New Year's Eve, to discover that its owner had converted it into a barricaded strong point during their absence. As they approached, the constables were greeted with a burst of gunfire which wounded one of their number seriously. As the weather was 45° below, his mate decided to make the long trek back to Aklavik, with King, the wounded man, riding the dogsled.

Certainly down to here the story is not in the Hollywood get-your-man tradition, the Mounties apparently having acted like policemen with normal human reflexes. Maybe the Mounties were keeping the need for climax in mind. Or maybe they were just ordinary mortals more courageous than most and certainly tougher, but still human. They came back to the cabin on January 9 with a force of seven men and forty-two dogs, using guns and dynamite, but dislodging no Johnson. This time intense cold drove them back, and when they returned their quarry had fled, which gives the Hollywood tradition still another beating. Three weeks later, Johnson's tracks were picked up in the mountains, high up the river, where he was believed to be hiding in the woods. The constables caught up with him this time. Shooting started and a Mountie, named Millen, was killed. At which point Wop May flew into the story.

May was at McMurray when the Mounted Police sent out a call for a plane. May took off. Weather slowed him as he flew north and three days were spent in making Aklavik. On the fourth morning he flew out to the police camp and landed on a frozen lake, close to the spot where Millen had been killed. Johnson meanwhile had struck over the mountains into the Yukon. The Mounties by this time were out for blood.

The pilot searched inland, flying just over the treetops, and picked up the Mad Trapper's trail along the Barrier River. After reporting back to the searchers he flew Millen's body out to Aklavik and returned with food supplies for the police party.

Several days passed during which May flew up and down the country searching for the fleeing desperado. He picked up the true trail again on February 14 and dropped red streamers to guide the ground force. But Johnson had obliterated his tracks, a little way ahead, by mingling them with those of a herd of migrant caribou and another three days elapsed before May picked his trail up again. This time it was clear and traceable, leading the searchers toward the Eagle River, deep in the interior. May flew back and forth, showing the way, as the police party converged on the area. Then Sergeant H. F. Hersey, driving the leading dog team, swung around a bend straight into the muzzle of Johnson's gun. Johnson fired, Hersey fell, seriously wounded. The others closed in for the kill – and kill it was.

The first problem was Hersey, Hersey and weather which had closed down into the treetops. Dare May risk flying him out? The pilot suggested that was precisely what he was there for. The wounded constable was laid carefully on the cabin floor, where Jack Bowen, May's flight engineer, tended him while May skimmed the trees and flew through the river bottoms, much of the way not ten feet off the ice, and brought his man to hospital at Aklavik, where the doctors said another half hour would have written off Hersey's chances of survival.

One more chore remained. May flew back to the police camp, brought Johnson's body out to Aklavik for burial, then flew that of Constable Millen out to McMurray for shipment home. When the surprised pilot reached Edmonton, it was to find himself a continent-wide headline figure and a local celebrity.

Such occasions bring brief fame, but the services rendered to the Mackenzie Northwest by a May or a Gilbert, a Dickins or a Brintnell, all the early bush pilots who flew the wilderness airways, are not measured in newspaper headlines. Their contribution was the smashing of the barriers that barred the way into an empire rich as Croesus, and, until they came, the greatest untapped source of natural resources in the Western Hemisphere.

LaBine spent the summer of 1931 with a prospecting party on the ground surrounding his original discovery. By fall a large area had been staked. The discovery itself had been trenched and thoroughly sampled. The mining world began to whisper that great events were afoot behind the curtain of the north. Around the streets where mines are measured in terms of marks on a strip of ticker tape, the scoffers, as always, were busy. Who ever heard of bringing in a mine clear up on the Arctic Circle? Look at the transport problem! Figure how much it would cost to ferry a ton of material, machinery, drills, supplies, over that fantastic water route! But there were others, men of adventurous spirit, men like Jim Richardson, who know that continents are opened to man by people who undertake the jobs that can't be done. So LaBine was able to press on with his pitchblende mine.

News of the discovery touched off a wild rush into the country. In 1932 prospectors and staking parties in droves roamed the coast of the lake. Overnight, Cameron Bay, just around the corner from LaBine's strike, had a post office, a restaurant, a hotel, a number of hastily built cabins, radio service to the outside world. In 1932 Richardson's planes packed 120,000 pounds of freight in to Great Bear alone, and down at Fort Norman the oilmen broke open a capped well to provide fuel for the diesel engines LaBine had hauled in over the Mackenzie and up the Bear River to the lake. The boom was on. Strangely enough, in the end it produced little but LaBine's amazing discovery; some fine showings of silver, but no other major occurrence of pitchblende. Had LaBine found the only deposit of radium-bearing rock in the whole country? To that nobody would know the answer for long years to come.

Meanwhile LaBine plowed ahead against almost insuperable odds. Freighting costs over the long water trail were, as the Jeremiahs had predicted, ruinous to a man pouring out capital and not likely to realize a dollar in earnings for two or three years, if then, until he could bring his discovery into production. Water transport into the country was still in the hands of the Bay Company, its sole remaining monopoly.

Pursuing a tactic employed in the days when fur was king and the fur barons were kingmakers, the Bay held freight rates at extortionate heights, refused to cut them to levels which would have made the carriage of heavy machinery feasible. Clearly the furmen wanted no part of resource developers in their country. If prohibitive freight rates would discourage them, so much the better. It was a shortsighted policy, for in the end it drove the Bay out of river shipping and made them suppliant customers of the crown. What matters here is that it was a policy that forced LaBine, against his will, into river transportation. Despairing of help from the fur barons, he flew east, ordered steel ships from a province of Quebec shipyard, freighted them almost three thousand miles from eastern Canada to Fort McMurray, on the banks of the Athabaska, and launched the north's first diesel-driven freighters, *Radium King* and *Radium Queen*, hauling ships for the northern run across the portage at Fitzgerald on rollers and launching them again at Fort Smith.

At the tumultuous rapids on the Bear River roustabouts manhandled heavy machinery, boilers, huge drilling equipment, everything that goes into the digging of a mine and, later, the erection of a concentration plant around the falls. At the rim of the lake, where Franklin built his fort, tugs and barges took over for the haul to the minesite, 160 miles away as the crow flies and more than 200 when the coast is followed, as is custom, for this is treacherous water and it is better to arrive than to lose cargo, barges, and tug in one of the quick gales that can whip the lake into fury almost without notice.

Toward the middle 1930's LaBine brought his radium mine into production, established a refinery in eastern Canada and began to feed the precious substance into the world market. The first effect of the new mine's output was to smash the monopoly prices established by the Belgian cartel, with mines in the Congo. But for all the glamour of radium, the mine prospered only indifferently. Basic revenues came not from radium but from uranium salts, widely used in the ceramics industry in Europe. Hitler knocked out LaBine's customers when his panzers swept west in 1940, and in doing so he shut down a mine more than five thousand miles away, astride the Arctic Circle in Canada.

Other minor strikes and other rushes followed on the heels of the Bear Lake boom. That at Goldfields, of fabulous proportions while it lasted, ended with no great discoveries, but one major operation based on huge low-grade gold-bearing ore bodies. In the Slave Lake country things were different, however. The boom that began with the discovery of gold in Yellowknife Bay in 1935 has never ended, though it has passed through occasional lulls, but has fanned out through the heavily mineralized country north of the lake to points more than a hundred miles north and east.

What is now the town of Yellowknife has been the center of this activity from the time of the rich Burwash strike in 1935. A year later, when the writer flew in from Goldfields with Brintnell, on the way to LaBine's Eldorado, a handful of prospectors' cabins stood in clusters around the shore, each group representing housing accommodation, cookery, storehouse, and machine shop for a small working party stripping gold veins. Yellowknife was not then a town, though in its enthusiastic moments, as on Saturday nights, it liked to speak of itself as such. A decade later, immediately after the end of World War II, when the writer next saw it, again en route to Eldorado to revisit the mine that shook the world at Hiroshima, it had become a thriving modern-minded community, an organized town, with its own air-strip, on which airliners from Edmonton touch down every afternoon, flown in by uniformed pilots, and tended by pretty stewardesses, serving hot meals over the western wilderness. Yellowknife in 1945 had a steam-heated hotel, running water, electric light, nine restaurants, a movie theater, a dozen or more retail stores, wholesalers' warehouses to supply the surrounding country, taxicabs, three banks, churches, barbershops, a school, a beauty parlor, a weekly newspaper, doctors, dentists, lawyers, local plane service to outlying lakes where mineral exploration was in progress, and a town jail. More than three thousand people were living in the district. Most important of all, seven producing gold mines had been established, with others of promise on the way.

Postwar Yellowknife was an enthusiastic town, promising itself a population of twenty-five thousand, a town with a future and no past. Three water-transport companies competed for summer freights. In winter tractor-drawn trains of sleighs, complete with caboose for crew living quarters, crossed the big lake to the mouth of the Hay River and ground their way laboriously south over a winter road through the wilderness to reach the railroad on the Peace. Townsfolk were then agitating for an all-year railroad down to the mouth of the Hay, and talked of laying tracks across the ice of Slave Lake in winter, to be taken up come spring. Why not? they asked. The Russians used to do it on Lake Baikal, clear back in the days of the Russo-Japanese War. So why wouldn't it work in the Territories?

This is the spirit that keys the new age that has dawned in the north, since the first airplane flew down to Resolution and Simpson, with Dickins at the controls. For two hundred fifty years the country slept. In the ten years that followed the coming of its first airplane, it leaped forward from the feudal age to that of the machine, of which the mining man is a straight-in-line product. The mining man wants the amenities of his age and he wants them in a hurry, not a quarter of a century after he founds a new settlement. The furman wants to be let alone to trade with the Indian on terms not greatly at variance from those which Alexander Mackenzie introduced to the great river in 1789. To ask two such communities to live side by side in amity is asking the future to mate with the past.

The transformation was not wrought without its crises and its agonies. Such stories come through in grim tales of mercy and ambulance flights. The record for 1932 alone contained thirty-eight such calls on the planes, to bring out men with frozen limbs, snow blindness, ptomaine, appendicitis, victims of accidents in the bush. There were fatal plane crashes in 1932 and 1933, each time in the rugged country north of Great Slave Lake, where the winter weather is anybody's guess from hour to hour.

The tales of the mercy flights are without end. One will suffice to show the nature of the feats the bush pilots performed and serve to introduce another heroic figure from the ranks of the empire builders. Matt Berry was a broad-beamed little man in middle age long before he quit the air, with more the look of the small-town businessman than that of the romantic, devil-may-care wilderness flier. But Berry could be devil-may-care and careful at the same moment, and if you knew him and had flown with him north of the treeline to the Arctic coast, and out over the snow-covered, frozen sea to Reid Island or Cambridge Bay with the mail, you knew that here was a man who flew by instinct, in country in which, in winter, it is impossible for the tyro to tell where land ends and sea begins in the vast whiteness that merges into the dim horizon.

Early in December, darkest hour of the arctic winter, when daylight is almost unknown, a radio call came south for a plane for urgent rescue. Unless help could be sent at once, the Roman Catholic coadjutor bishop of the Mackenzie District, Monsignor Falaize, with two priests and three Eskimo children, faced certain starvation at an isolated mission on the Hornaday River, near the Arctic coast, 200 miles north of Great Bear Lake, in country into which no plane had ever been flown.

Berry took off and pressed through. After crossing the Circle into terra incognita, he flew on in a dim dusk in which it was impossible to read the instrument panel. Two days out from railhead, he reached the mission, landing on rough ice in almost total darkness. That night the bishop and his party ate their first food in several days and told Berry the story of their travail.

They had been working up the coast of Coronation Gulf, trying to reach Burnside Harbor, but somewhere not far from the mouth of the Coppermine their mission schooner, *Our Lady of Lourdes*, had been frozen fast and they had abandoned her. Reaching shore, they had traveled 75 miles across country on a compass course, making for the deserted Hornaday mission, where they hoped to find food. The expected cache was gone. Their sled dogs died of starvation. The nearest fishing was 30 miles away. Letty Harbor, where they knew they could find succor, was 50 miles to the north, and they had nothing with which to sustain even one member of their party for such a journey.

Next morning Berry flew to the coast in foul weather to bring food from Letty Harbor. That was the first consideration. The party then waited for a break in the weather to begin their journey south. On the fifth day they got off and Berry set a compass course in the dusk for Aklavik, but over Darnley Bay, according to Berry's log, "it became too dark to follow anything" and he turned back to the mission while he could still glimpse the dim outline of the land under his skis. They waited another five days. On the sixth they flew as far as Bishop Lake, where they landed with "visibility nil and light terrible."

That night, while the bishop and his party slept in the plane, body to body to share each other's warmth, Berry and his flight engineer nursed an engine, threatened by a temperature of 40° below, draining it and keeping the oil heated, to be poured in again in the morning. Next day they got off the lake at 10:45 and set a course for Aklavik. This time they made it.

From Aklavik, Berry followed the Mackenzie up to Good Hope, in a temperature of 60° below, and flew on to Norman, where he cut west for Great Bear Lake and the Eldorado, but found a landing impossible in heavy fog which closed in as they crossed the big lake. They pushed on and reached Resolution, and the next day Fort Smith, 35 miles north of which settlement Berry spotted a plane down on a slough in the back country. He unloaded his passengers at Smith, took off, landed on the slough and picked up pilot Gil MacLaren and his engineer, flying for a competitor line, Mackenzie Airways, which Leigh Brintnell had organized the previous year. Berry brought his stranded competitors in. Next morning the party set out again, with Edmonton as the day's destination. They reached McMurray during the morning and pushed south until snow drove them down and they landed in a field 80 miles north of the Alberta capital. In making the landing a ski pedestal snapped.

While his charges slept, Berry and his engineer improvised a new pedestal. Berry took off solo from the field to test it, while the others mushed overland to a small lake, invisible in the previous day's snowstorm. The new pedestal survived the strain of landing and the party climbed aboard again. An hour later they were in town and another mercy flight was over.

Such tales are without end. Mercy flights and rescue flights. Flights to bring out murderers and to carry the Mounties to the scenes of the north's rare crimes of violence. Flights of exploration: Gilbert's to the Magnetic Pole, Dickins's first flight across the Barrens, over the trail first blazed by men like Back. Flights to find missing fellow pilots, as when Berry pulled Con Farrell out of the Barrens, after an 11-day sit-down with a burned-out engine. And behind it all was the steady drive of day in and day out flying which, as the gold-seeking frontier busters increased in numbers, piled up mounds of freight at railhead for movement north and brought into the country a steadily increasing flow of fortune seekers. That was the bread-and-butter flying. They'd take you anywhere, the wilderness pilots, not merely to the settled areas but to any back-country lake you could name. They'd drop a man and his gear and make a date to come back, maybe two months thence, and if the weather was flyable that day you could count on your pilot to be at the rendezvous waiting. They were the corps d'élite of bush-country aviation, and they opened an empire that had been closed to all but a few men since the beginnings of time. Canada has an award, the McKee Trophy, given each year for outstanding contribution to aviation. Among its winners are

Wop May, Punch Dickins, Matt Berry, and Walter Gilbert, in tribute to the roles they played in opening the Mackenzie Northwest.

When war came in 1939 and the threat of attack over the top of the world became acute and actual, it was to the men who had pioneered the air routes that military leaders turned for the know-how that gave their armies aerial lines of communication in the nick of time. Had it not been for the armed forces North America was able to throw against the enemy, over lines pioneered by the bush fliers, a Japanese attack conceivably might have reached into the mainland and have led to land-based air raids launched against midwestern cities from North American soil. These are the terms in which the contribution of the new empire builders must be measured, a few to whom a continent owes much.

CHAPTER NINE

New Northwest Passage

AT MIDNIGHT on September 9th, 1939, the Canadian government, with the almost unanimous sanction of Parliament, requested his Majesty, King George VI (who is King of Canada in a role completely separated from his status as King of England), to inform the German government that a state of war existed as of that hour between the people of Canada and the Third Reich.

In the ensuing excitement the eyes of the public were focused on Europe as men were recruited, trained and shipped across the Atlantic, in precisely the same manner in which their fathers had gone away to war twenty-five years earlier. But behind the scenes in Ottawa and Washington small groups of men met in deep secrecy to reach vital decisions involving war, not in another hemisphere but in our own.

Suppose an attack were to be launched on North America over the roof of the world, what then? How could the continent defend itself, repel invasion, or prevent an enemy from establishing beachheads on the soil of the Western Hemisphere, from which to launch aerial attacks on industrial centers in the middle west? Alaska was virtually cut off from its parent, the United States, excepting the tortuous water route up the Pacific coast from Seattle. The Yukon was in similar case in respect to Canada. Between Edmonton and the Far North not an airdrome existed. Beyond the Peace River country in the west and Fort McMurray on the Athabaska, no railway penetrated the wilderness, no highway crossed the interior. North America, in short, possessed no overland communications joining its populated areas, centers of production, and military bases with the region where, if attacked, the continent must be defended.

The first step was Canada's alone, in part because the United States was not then at war, in part because Canada was determined to conserve its integrity and sovereignty and, therefore, to finance all military activities on its own soil. The writer may be excused by the American reader if he emphasizes this point, because misconception has existed in the United States as to the nature of wartime activities in the Far Northwest, and the belief has been current that it was the Army of the United States that created the defense system against the Japanese in the vast continental hinterland. Later, after being attacked in Hawaii, the United States would balance the books by thrusting a highway across the wilderness to reach its own soil in Alaska. But during 1940 and 1941 the burden was Canada's and during that time the northern dominion established in the almost impenetrable wilderness the chain of military airdromes known as the Northwest Staging Route.

Over that route went the first urgently required supply for defense of the continent. Over it, later, poured a stream of Lend-Lease aircraft to our Russian allies. Canada's series of airdromes in the wilderness became, in urgent fact, the lifeline of North America's western defense and, when defense changed to attack, a new northwest passage to the Orient by air.

Canada's old-time bush pilots had blazed the trail. Years before, one of these, Grant McConachie, began flying into the Peace River country of northern Alberta. He flew mining machinery and supplies into the northern settlements and came out loaded with fish from northern lakes for city markets. He flew farther in and established a regular service from Edmonton to Whitehorse. His company, Yukon Southern Air Transport, carried the mail and operated regular passenger and express service in twin-engined all-metal aircraft from Edmonton into the Yukon.

In 1939 ground-survey parties had been sent in, and the new route was recommended to the Joint Defense Board as a channel to Alaska. The plane called for main fields to be established at Grand Prairie, Fort St. John, Fort Nelson, Watson Lake, and Whitehorse. The first two sites were near enough to railways to eliminate any major transportation problem in the carrying in of construction materials. The Whitehorse site also was close to the rails. But Fort Nelson was 300 miles north of the end of steel and Watson Lake was not even a pinpoint on the map of the vast northwest hinterland. Now the engineers closed in on the wilderness.

Parties were flown to Fort Nelson during the winter of 1940-41 and started laying out the airdrome. Engineers and clearing crews were also flown to Watson Lake, just north of the Yukon border and about 240 miles by air from Fort Nelson.

Loaded with building materials for Fort Nelson, a tractor train started out from Dawson Creek in two sections on February 9, 1941. Leading the first was a bulldozer fitted with steel blades capable of mowing down trees and slicing the tops off knolls by sheer brute force. Behind it moved another tractor hauling several wooden huts on sleighs, containing a combination kitchen and dining room, staff quarters, and two bunkhouses for the crew of twenty-eight men. Behind this section came another consisting of a blacksmith shop and five sleighloads of heavy freight, hauled by a big tractor.

Normally, the train moved ahead four times a day, in the morning, at noon, at evening, and again at midnight. Hauling the cabooses as far as possible, the big caterpillar then doubled back for the freight and blacksmith shop. In many cases, due to steep grades, bare ground, or other forms of heavy pulling, the sleds had to be taken one or two at a time. Regardless of how much this general system varied, it was always planned to have the kitchen up to the bulldozer at mealtimes.

The large trailbuilder necessarily was always in the lead. In order to allow passage of the broad sleighs, more than one cut often was necessary to make each leg of road. Thus, in the morning, it would move ahead for two or three hours, then backtrack to widen and smooth. In the meantime the rest of the train was being brought up, until at noon the two met. After dinner, the bulldozer continued making road and the remainder of the train came up to meet the first section. If the type of country permitted, the smaller bulldozer did widening and general conditioning in the rear of its mate. Using either method, from one to two miles were averaged per move and on occasion as much as ten miles were covered in twenty-four hours. Far ahead of the bulldozer, trail blazers, first with horses and later using dog teams, chose the route. This followed, generally, an old freight trail to a trading post on the Sikanni River, but the trail had to be widened to accommodate the larger sleighs, and then smoothed and filled, as trucks were to follow as well. It was, at best, a beginning.

North of the Sikanni, only a toboggan or pack trail existed. The train deserted this almost entirely until within a short distance of Fort Nelson, due to its crooked course.

A tremendous project remained behind the bulldozers.

More than six hundred tons of general freight, including everything conceivable with which to begin construction of the Fort Nelson airport and for the needs of the construction crew, had been loaded onto sleighs and trucks. Trains 3 and 4, hauled by two large caterpillars, including a bulldozer, and followed by twenty trucks, shuttled freight, first to a cache at Nig Creek and then to the north bank of the Sikanni. Here everything was assembled for the final pull into Nelson. From here in the going was expected to be tough. It was.

The hardships endured by the train crews and truck drivers were tremendous. Yet scarcely had the leading train reached the Sikanni than the heavy freight began to arrive. As the road moved into the last stretch, materials were piling up on the north bank, approximately one hundred miles south of Nelson, waiting for road for the last pull into Fort Nelson.

As the trains were en route, nature took a hand. Early-breaking spring descended on the muskeg. What had been a frost-bottomed road became a quagmire, the sticky fingers of which clung tenaciously to caterpillar tracks and dragged trucks down to their hubs. The trail itself was churned into a river of mud and water, forcing the trucks to turn back at the Sikanni, while the remaining cars and engines, assisted by those of Train No. 1, worked day and night to bring through all possible freight before breakup. Gradually the material arrived and almost two hundred tons were piled at the new airport site before travel became impossible. River ice broke and went out, the trail became a series of roaring torrents, caused by the rapid melting of four to five feet of snow. The earliest breakup on record had disrupted carefully laid plans.

Breakdowns delayed the train. Three times in the first week repairs held up progress for a total of twenty-seven hours. On the last occasion the severity of the break necessitated a return, with small tractor and caboose, to the blacksmith shop at Fort St. John. Thus seven days found the train only 35 miles beyond the last settlement, Murdale, a poor beginning.

The second week brought better results. The size of the timber made it possible for the bulldozer to work satisfactorily and lack of thick underbrush permitted broken trees to be pushed completely from the road. Grades, too, were favorable. February 23 found Trains 1 and 2 camped alongside a deserted fur-trade settlement at Nig Creek.

The old log buildings, once the nucleus of a thriving trade, stood in a small clearing, monuments to a bygone day, their weather-beaten sides and collapsing roofs silently hostile to the encroachment of the new era of the diesel tractor and modern trailbuilding skill.

The crossing of Nig Creek presented an obstacle common to practically all the rivers and creeks. The old trail, accommodating only horse and dog teams, had followed the path of least resistance. Winding approaches added unnecessary miles and made turns impossible for a connected train of wide sleighs. In many cases the trail lay on the river ice. Hence it became necessary to search for more convenient grades, to cover a minimum number of miles, to cut the amount of ice travel, and to "fill" at the creek crossings. At such points the bulldozer pushed large timbers out on the ice and piled smaller logs, brush, earth and any available material into the apertures until a 10-foot roadway had been built.

Again and again such fills were made. Farther on, the Pine-to-Conroy-Rivers section entailed no less than four such projects, each presenting individual peculiarities. Up and down the bank went the bulldozer, cutting in where the trail lay across the faces of hills, filling holes and washouts and tearing at the frozen ground to throw it into dry creek beds and onto stretches of ice. At one place, fill was put into a 50-foot river so that a level road lay from one flat to the other. Across this the train pulled two sleds at a time over what a few hours before had been a river valley under four feet of snow and completely filled with trees and underbrush. Now, a 20-foot highway led from the stream for half a mile, up a 10 per cent rise. But at the Sikanni River crossing fill was out of the question and the only alternative was to use the ice. Cautiously the bulldozer started across the river, feeling its way. Near the opposite shore the ice suddenly gave way. Without warning the massive machine slipped through to the river bottom and began to sink in soft mud. Timbers jammed under the tracks only slowed its descent. The other engines, pulling on the ends of cables, failed to have any effect, as quicksand had closed about the tracks.

Working in the ice-cold water, men with shovels dug

at the river bottom, slowly releasing the caterpillar tracks from the quicksand's viselike grip. At the same time the two tractors, their own tracks slipping and skidding on ice, held cables taut, as inch by inch and foot by foot the dead weight began to move, as hour succeeded hour. All night the work continued. Gradually the light of gas lanterns faded before approaching day, as the quicksand's death clutch was broken and the trailbuilder rose higher and higher from the water.

Frozen mud caked the tractor tracks into a solid mass. Water had penetrated the engine, necessitating complete cleaning and drying of the lubricating and ignition systems. Against time and bitter weather, the repair job went on. The Nelson Trail must be pushed to journey's end.

The last lap had begun. The Sikanni became a memory. Eyes turned wearily to the last hundred miles of muskeg, a country of light timber and tangled underbrush. If trucks were ever to come through, it meant working and reworking the newly made road repeatedly, and for that there was neither sufficient time nor equipment.

Here the vegetation, for the most part, consisted of small spruce, birch and poplar saplings and 30-foot willows, a rubbery mass so thick as to make impossible the pushing aside of rubble from the road. Hence an extra width had to be made; the center for passage of the train and the sides to accommodate slash thrown aside by the bulldozer blades. No longer did the trailbuilder pit its strength against large trees and send them reeling away from the road. Now it must rip and tear at stubborn saplings that bent before the blade, leaving five or six feet of peeled willow with roots still intact after the bulldozer had passed. Moss and lichen, shaken loose and falling from the undergrowth, were drawn against the radiator, so that for days it became necessary to stop at two-hour intervals to remove the protecting grille and clean away this accumulation, to permit air to reach the overheated engine.

Day by day the distance increased, however, until March 27 found the train camped beside the Muskwa River, on the threshold of the Fort Nelson site. Four miles ahead lay the end of the trail, but one more river must be crossed and warm weather had wrought havoc with the ice. A careful check was made and finally a crossing was chosen which seemed to offer good support.

At first an attempt was made to pull the bulldozer across on two large skids, but this proved impracticable. Finally it winched itself across (the winch cable was anchored to the opposite bank and, with tracks running free, the machine began reeling in). Steadily the distance lessened, 40 feet, 20, 15 – and then misfortune struck. The ice gave under the big machine's weight.

The tractor staggered to one side, then to the other. Frantically the driver tried to get his tracks in motion, but before they could respond fifteen tons of metal had plunged to the bottom. Five feet of water closed around its sides, flywheel and fan belt sent showers over the cab and with a sputter the engine, strangled by ice-cold water, died. Was the game up, with the end of the trail in sight?

The driver extricated himself from the cabin, expressing in eloquent terms his opinion of the combination of heavy machinery and thin ice. Immediately the task of cutting a channel to shallow water began. All night and most of the next day were spent in cutting free large blocks of ice. These the smaller engine pulled from the water, then it dragged the bulldozer up the slope at the end of a cable, until by three o'clock the following afternoon it was in sufficiently shallow water to start work on the engine. Once again overhaul was necessary to remove ice from the machine. Hours later, under its own power again, the great engine of war-in-the-bush began to climb to solid footing. Meanwhile a new course had been selected, and the train moved over the river flats on the evening of March 29, to spend most of the 30th undergoing repairs. On the last day of March the leading train climbed its last hill and camped beside the future Nelson Airport, forty-eight days out from Dawson Creek. The goal had been reached and, although it was not the intent, the Alaska Highway had been started.

Equipment and supplies for the Watson Lake field could not be moved in until the waterways opened in the spring. A stern-wheel steamer, two tunnel-bottomed boats, and a number of barges, 55 feet long, were built in Vancouver for the job. These and eight hundred tons of material were shipped up the Pacific coast to Wrangel, Alaska, in the spring of 1941, then were freighted 163 miles up the Stikine River to Telegraph Creek, and hauled 72 miles by truck over a tote road to Dease Lake. Here boats and barges were assembled and, supplemented by local water conveyances, were used to freight the supplies 193 miles down the Dease and Liard rivers. The final stage into Watson involved bulldozing a road 25 miles from the Liard to the lake. Meanwhile aircraft had taken in a sawmill and planing mill, to prepare lumber for buildings.

By the time the Japs struck at Pearl Harbor a chain of outsize, but still rough, military airdromes was in service from Edmonton and Vancouver to Whitehorse in the Yukon and on to Alaska on either side the Rocky Mountain barrier. North of Edmonton, the easterly route ran to Whitehorse, with airdromes at Grande Prairie, Fort St. John, Fort Nelson, and Watson Lake. The route on the western side of the Rockies ran north from Vancouver and Kamloops to Williams Lake and Prince George. Another field had been built at Smithers, near Prince Rupert. Unfinished and unequipped with beams and other safe-flying devices though they were, the fields nevertheless could be used in emergency. The lines of communication were taking shape.

When finished, the Staging Route fields were equipped with landing strips 4,000 feet long by 500 feet wide, with full lighting equipment for night flying, and a radio beam system to guide aircraft over the route, night and day, in all kinds of weather.

Construction completed, grading units, bulldozers, plows, rollers, and 2-ton truck equipment remained in the interior to build intermediate emergency fields, between Fort St. John and Fort Nelson, between Fort Nelson and

Watson Lake, and between Watson Lake and Whitehorse. Canada had opened the new Northwest Passage. For once a democratic nation had thought a little way ahead.

In the early days when the Staging Route, to which the U. S. Army gave the code name Alsib, a combination of Alaska-Siberia, was flown, military fliers, unaccustomed to the bush, labeled it the world's toughest airway. Veterans called it rougher than the Himalayan Hump into China, and it was five times the Hump in length. Maps between airdromes were inaccurate, for much of the region was not merely uninhabited but had never been flown until the dromes went in. Lakes and mountains suddenly turned up at points far from where the map said they should be. Melting snow in spring created "false lakes" to throw a pilot's judgment out of kilter. In summer forest fires blacked out landing fields as tight as any fog could. In winter blowing snow obscured runways.

These were problems with which the bush pilots of the northwest had coped ever since they had gone into the Territories, but to most of those who came ferrying military aircraft the conditions encountered were of a character for which they had little if any training. In the air the youngsters in uniform battled snow-created static conditions never met south of the border, which played strange tricks with radio beams. Mountains and ore deposits turned straight-line beams into dog-legs. Winter weather changed from fair to blizzard in ten minutes' flying time, shutting down to zero-zero fields which had been wide open when the pilot began his approach. Winds blew up to 110 miles an hour. In the beginning only 12 weather stations served the whole outlying area, far out into the Pacific, when, according to estimates, 300 were needed to keep the men on Alsib posted on what to expect. By the end of the war they had 125 and things were looking up, but they still had far from sufficient meteorological information to assure secure flight.

On the ground, the northern lights played hob with interfield teletype systems. When Aurora Borealis was well behaved, bears chewed and cut the lines.

Men were killed. Twenty American pilots and many more military passengers were lost before the end of 1944. Eleven men died in a single day during the first winter operations, nine of them passengers and crew aboard an army transport believed to have hit a mountain while flying in a cloudbank: "hard center clouds" is the old-time bush fliers' phrase for it. Two others were killed when they crashed in a blizzard that night, while trying to make their way into Watson Lake. The two survivors of this crackup were not found until sixteen days later.

But for the know-how that the bush fliers freely gave to the military newcomers it could have been much worse. From May and others, the army boys learned much of the score on winter flying, were given tips by the hundred which could have been found nowhere in books or on maps — tips on where and how to land, on recognizable weather signs, how to get out of the strangely peculiar jams that can come at a pilot out of the blue, down north, what spare parts to be sure to carry, how to keep alive if grounded, what to wear in subzero climates. It was Canadian Airways' Tommy Siers who developed an oil-dilution process which solved the cold-weather starting problems of outsize military motors and engraved his name on the McKee Trophy. The bush fliers, many of them in Canadian uniforms by this time, had brought their great know-how to the aid of their young allies from south of the border.

By the end of 1944, six thousand American-made aircraft had been delivered to the Russians over the Alsib line. Assembly point in the United States was at Great Falls, Montana, whence planes were flown north to Edmonton and over the long hop-skip-jump into Fairbanks, Alaska, where Russian crews took over for the 6,000-mile onward passage to Moscow. The movement began in 1942, with departure of five bombers on August 31 from Great Falls. The next winter, the U. S. Army Air Force's first on the run, only fourteen aircraft were delivered into Fairbanks. Then the White House stepped in with clear-the-way priorities from assembly lines to collection points, and the door was opened to mass delivery. Planes for Russia were given first call, even over delivery to American and British fliers in the United Kingdom. Stalingrad must be saved if Russia was to be kept in the war. The bombers went through and Russian pilots gained mastery of the air from the Luftwaffe. A wilderness airline across northern Canada had turned possible defeat into victory, on a front 6,000 miles away.

The tremendous wartime activity in the Territories and

the Mackenzie Valley itself were to revolutionize postwar aviation in the country. The sudden decision to thrust a military road through the wilderness to Alaska and the building of the pipeline from the oil wells at Norman through the mountains to Pacific tidewater necessitated a rapid-transit system within the country itself, exclusive of the aerial delivery shuttle over the Staging Route, to support and supply highway and oil developments.

To serve highway builders and pipeline engineers landing fields were constructed throughout the areas involved. Fields went in near McMurray and at Smith, with an emergency strip halfway down the line at Embarras, close to the Athabaska delta. Resolution and Hay River became stopping points on Great Slave Lake. On the Mackenzie itself fields were constructed at Mills Lake (near Fort Providence), Wrigley, Norman Wells, and just across the river, at headquarters for the Canol Project. Soon, for the first time, wheel-equipped planes were winging over the big river. Today they carry virtually all the passenger and express traffic of the main lines, as far as Norman on the river and to Yellowknife in the east, where a field has been built since the war. Thus war activity has sharply changed airplane disposition in the north. Today twin-engined passenger liners carry all main-line commercial business. The smaller, single-engined pontoon- or ski-shod planes operate primarily on local charter services, out of such central points as Yellowknife into the hundreds of lakes that are still the airdromes of the back country.

For years before World War II, Americans in Alaska and Canadians in the Yukon had been agitating for a road to connect their isolated communities with the United States and Canada proper. Nobody in Washington or Ottawa would listen. But when Pearl Harbor brought political leaders and soldiers alive to the urgent threat to America from an enemy based in the Pacific, the road was laid on overnight, to take the supply burden off an already overloaded air route and to provide ground communication for American and Canadian troops engaged in a phase of the war which had brought the enemy within steppingstone distance of the North American continent, at Kiska and Attu in the Aleutians. Moreover, if the airfields of the Staging Route were to be secure, a connecting highway was needed, with exits to south and north. So Alaska and the Yukon were given their road, 1,671 miles long, 1,200 miles of it through the heart of the empty Canadian hinterland.

The highway venture was one of those arrangements which light up the close co-operation of the two countries that share title to the North American continent. Canada had built and paid for the airdromes, by this time primarily used by the American Army, although when undertaken by Canada its government had no means of knowing that soon it would have a war partner in its next-door neighbor. Now the United States offered to even the score by bearing the expense of a road, using U. S. Army Engineers to build it. Canada consented, on the understanding that at the end of hostilities that part of the highway which traverses Canadian soil should revert to Canada and be patrolled by Canadian, not American, forces. The United States agreed and Uncle Sam moved in.

During the winter of 1941-42 engineers and equipment began to pour into the main operational base in the Alberta capital, in such numbers that wags christened the city "Edmonton, U. S. A." The rachitic railway line that pushes out into the Peace River country was choked with the forward movement of supply. Flatcars, loaded with bulldozers, graders, steam shovels, all manner of outsize roadbuilding gear, rattled down the line to Grande Prairie, whence equipment moved down the existing motor road to Dawson Creek, where an airfield was built in a hurry to serve the project.

In March, 1942, construction gangs set out from Dawson Creek to break a highway through the wilderness, already roughly broken as far as Fort Nelson by the airport builders. Allowing for regional detours, its primary line was due northwest, to Fort St. John, thence to the Staging Route field at Fort Nelson, on the headwaters of the Liard River and to Watson Lake. Farther on, the road hit the long narrow lake south of Teslin and followed its shore north, thence swinging west and jogging through the hills into Whitehorse, with a cut-off to nearby Carcross. From Whitehorse the road was driven through the mountains again, to journey's end at Fairbanks.

Eight months after the first bulldozer bit dirt the Alaska Military Highway was open for business throughout its length. Truck convoys loaded with supply, ammunition, all manner of military equipment, rolled north through clouds of blinding, throat-searing dust, wallowed in snow through the eternal winter. Food, fuel, and repair depots were established along the line. A highway which carries through the bush for hundreds of miles without vestige of human habitation, excepting its own patrol stations, had been driven through. Without question, it was one of the great engineering feats of a war in which engineering miracles were the common currency of the drive for victory. So in war the people of Alaska and the Yukon secured their connecting link with the outside, a link for which they might have waited decades in a world at peace.

Meanwhile great and secret events were afoot on the rocky rim of Great Bear Lake.

On a morning early in 1942 Gilbert LaBine took an urgent long-distance call in his skyscraper Toronto office, which summoned him to the desk of the Honorable C. D. Howe, Canada's wartime minister of munitions and supply. Later that day, across Howe's desk, in the eyesore buildings where Canada did the greatest buying and supply control job in the nation's history, the minister asked how quickly LaBine could reopen his mine and begin uranium production.

LaBine fished a used envelope out of a pocket and began to figure, while he asked pertinent questions. Where could he get miners? National Selective Service would attend to that on a top-rank priority basis. What about supplies, tools, machinery for the enlarged program Howe wanted? What

about transport? Just holler, the minister said. Holler for anything you want and we'll get it for you in a hurry. But get cracking, Gilbert . . .

On these terms LaBine sped back to Toronto, his lips sealed even to his closest associates as to whatever part of the greatest secret men have ever kept inviolate had been given into his keeping by Howe. All his directors were told was that the Government wanted the mine reopened, but fast, its production increased to whatever the traffic would stand, that henceforth they'd have only one customer on the books, the crown . . . and, gentlemen, don't tell even your wives!

All this coincided with events in the field of science. Ottawa had started urgent research in the realm of atomic fission in 1940. When France fell, her scientists had escaped to Britain, to join forces with colleagues there. Through '40 and '41 progress had been made, and, as Howe and LaBine were making ready to provide uranium for the United Nations, Canadian and British scientists and here and there an expatriate from the Hun-overrun countries were gathering in Montreal to carry on research as a team. Dr. C. J. Mackenzie, president of Canada's National Research Council, headed up an imposing Canadian list. Munitions Minister Howe became the British-Canadian liaison officer with Washington. The key to the most secret and awe-inspiring military plan in history lay in the ice-lined ore bodies of Canada's Arctic Circle. If LaBine and his experts had not been able to give the scientists here and in the United States the makings of the precious and rare element U-235, Hiroshima might never have hit the headlines. No matter who gives or takes credit for what, Canada's contribution in the field of supply, over and above the work of her scientists, stands as a major factor in production of the Bomb, a contribution which can never be turned aside. Before the scientists began making bombs, they had to have U-235. And Great Bear Lake was the only place in America where uranium could be had in quantity.

The secrecy lid was clamped down, even while Eldorado was still a private operation, with thousands of shareholders, through 1943. If you were working at Eldorado, you didn't have much to say about your activities in your correspondence. The latchstring had been pulled inside as soon as the plant reopened and the airlines people were told not to accept passengers for the radium country, without first getting permission to bring them in. The mine, in turn, was receiving no guests unless they carried triple-A credentials, endorsed by the highest authority.

But there was no brass-hat fuss and furor about it. Officials kept their secret by being nonchalant, by attracting no public attention to the Big Hush. Then came February, 1944. The scientists had just said they had a chance to bring off the bang before the last innings of World War II, but they'd need as much U-235 as they could get. Right then the Canadian government, in the person of C. D. Howe, stepped abruptly in and took over, lock, stock and barrel. Eldorado became a crown company. All the uranium in Canada, found or to be found, became the property of the nation.

Howe offered no explanation. The government had taken over. That was all. When the occasional shareholder called it confiscation, nobody bothered to listen. Some of them were still growling in 1948, although the price they were paid for their shares represented a five-year high.

With the government in the driver's seat, Eldorado began to go. This wasn't just a case of digging out a hatful of professors and housing them in an ivory tower to carry on secret experiments leading to a destination privy to only a few. It wasn't simply a case of mining, concentrating, and refining enough of the supermaterial to make a few bombs. It was uranium for today and tomorrow. For peace as for war, for construction and not merely destruction.

The idea seems to have been spread around (due, no doubt, to the tricky nature of radium, on the one hand, and the awesome wallop packed by the atomic bomb, on the other) that Great Bear ore must be extremely difficult, if not dangerous, to mine. People ask if miners wear special equipment to protect them from radioactivity in the rock, if the ore can be handled with bare hands, if eyes require special protection.

The answer to each question is No. By the time pitchblende reaches the point at which its atoms can toss a city into the air it has been refined in such a degree that the finished product reaches a ratio to the original rock that is beyond any but a mathematician's computation. In that state it is dangerous to handle. But not in the raw.

Mining practices at Great Bear Lake are identical with those of any hard-rock operation. Working levels are 125 feet apart. From the shaft, at each level, a crosscut is driven over to meet the vein. There the cut turns and follows the strike of the ore. Along the way raises are sent up and stopes opened, a stope being a gradually increasing cavern cut to get at the ore between levels. From stopes the ore is sent down chutes into mine cars, to be hauled to the cage, hoisted up the shaft, and sent to the mill. The only difference between mining at Eldorado and, say, in Michigan is that at Bear Lake the job is wetter and colder.

The mining and refining processes of Eldorado are still top secret. Suffice it to say that at the minesite the ore is crushed, sorted, and put through the first of many processes leading toward its final destination. From the first of these it emerges as a coarse black sand that is shipped south, where it disappears under the veil of secrecy, until it emerges from the scientists' hands in whatever form it is finally used.

Not merely mining of established ore bodies but urgent exploration of favorable geology in the hope of discovering new sources was the task that faced the men of Eldorado. That search continues. In 1948 Canada, after barring private prospecting for pitchblende, opened the field and began to encourage the search. Experts are convinced more deposits can be found. The prospecting approach constitutes something new under the mining sun. Old-timers of the pick-and-dynamite school may mutter in their whiskers and growl that it couldn't be. But the hunters for radioactivity in the northern rocks have staged a revolution. They carry no picks, just scientific instruments.

At the outset, the scientists went into the universities looking for geology students anxious to do field work in the schools' off season, which is prospecting's open time down north. What they sought was not merely young men knowledgeable in rocks and the scientific approach, but persons possessed of self-reliance and the high IQ.

Knowledge of the bush? Not important, they said. You can't give a youth the scientific mind in a couple of weeks, but you *can* teach him how to live in the wilderness, a point of view calculated to shake any hoary bushrat with the palsies of scorn. Eldorado made up twenty two-men parties. Then they yanked a well-known engineer, Dick Murphy, bodily out of the army to direct exploration and brought Emil Walli, prewar Eldorado manager and an expert pitchblender, back from Africa as geophysical boss. A couple of Norsemen aircraft made their outfit mobile.

The new prospecting twist relegated the pick, the shovel, and the stick of dynamite to the second division. Each party carried as its primary equipment what is known around the places where scientists dwell as the Geiger Counter, a highly intricate mechanism containing a Geiger-Mueller tube, done up in portable form and worn like a light pack by the Atomic Age prospector.

When their airplane deposited a Geiger team on a lake adjacent to what geological maps revealed as favorable terrain, the party set off into the bush and, when it reached the region it had come to explore, one of the young men donned the pack, complete with headphones, and proceeded to prospect.

The technical explanation of what goes on at this point is that the instrument occupies its mind with picking up gamma rays from the rocks. Reducing the matter to lay terms, if there is radioactivity in the ground over which the team is working, the Geiger picks it up and proceeds to make strange noises into the prospector's earphones, the strength of the noises depending on the strength of the activity inside the earth.

The combination of scientific approach and airplane gave the Eldorado crews a mobility never before seen in the mine-hunting world. In pre-Geiger days pitchblende hunters might have worked all summer over a confined area and returned to civilization in the autumn with nothing to show for the summer's work but flourishing growths of beard. Under the Geiger-cum-aircraft method, parties were whisked into favorable geology, turned loose with specific instructions as to what particular zone to examine, and told to be back on the beach two days hence with their findings. The mining north had never heard of such goings-on.

By this combination of rapid transit with science, the pitchblende hunters moved over an area of roughly 35,000 square miles during two wartime prospecting seasons, seeking new ores against the day when Eldorado might peter out. The search took them from the home base south to the shores of Lake Athabaska, north to the mouth of the Coppermine River and the shores of Coronation Gulf, to seek radioactivity in rocks which experience at Eldorado, combined with geological maps, indicated as being favorable to their purpose.

The result was discovery of several likely places for the bringing in of new deposits, which Canada regards as no more than a start in the right direction, but still a chore which might have occupied a century of time under old-fashioned prospecting approaches, if it had been possible to do it at all. Later, in 1948, Canada raised the curtain that had concealed its search for pitchblende and invited all comers. With an ambitious atom program of its own – for peace, not war – as well as the neighboring United States to supply, Canada was preparing to go all out in the hunt for more pitchblende. LaBine's Eldorado has been called the Mine That Shook the World. In days to come perhaps it will be known as the mine that revolutionized man's approach to the uses of energy for the purposes of peace, not destruction – and the man in Eskimo parka and mukluk boots who flew to the Arctic Circle in 1929 as the citizen who touched off the fuse, not of a mining rush but of an industrial revolution.

One further item closes the book on wartime activities in the empire of the Mackenzie. It has been called the greatest blunder of the war years. It has been called all manner of things, all of them sharply critical.

Its name is Canol and it stemmed, like the Alaska Highway and the aerial Staging Route, from the urgent needs of defense against the Japanese. Oil is the key to modern war, the basic ration on which airplanes fly, trucks and tanks roll on the surface of the earth, and submarines cruise under the sea.

The Mackenzie Valley has oil. Someday, perhaps, Norman may be the center of one of the world's great producing fields. But in the world as it was in 1940, the wells on the Mackenzie were a minor manifestation, producing oil to feed the mines and the transport of the country. Then the Army of the United States moved in.

Overnight new wells were drilled. From the far bank of the river engineers and construction gangs set out to drive a pipeline hundreds of miles through almost impenetrable mountains to reach Pacific tidewater. Men who had run pipe across deserts under a broiling tropic sun now turned to a reprise of these tasks in temperatures of 60° below. As they pushed on into the Mackenzie Mountains, road crews constructed a highway to bring them supplies and food. The United States Army moved an oil refinery, holus-bolus, from Texas, up the full length of the Pacific coast, and set it up on northern tidewater. When press and radio announced that this great engineering feat had been completed and that arctic oil was ready to flow through the mountains to the sea, to supply Allied ships and planes with urgently needed fuel with which to fight the Japs, the military public relations peoples' handouts spoke in epic prose.

But with the return of peace people began to speculate

as to the future uses of the pipeline and the purposes to which its oil could be turned. Could Mackenzie River crudes be carried down the Pacific to supply seaboard cities? Could they be usd for industrial purposes in Alaska and the Yukon? Neither of these ideas was workable, the brass announced, explaining that the pipe had not been built as a commercial project, but as a defense measure, which could not be used with profit in time of peace. It had been an emergency business and it was being shut down. Uncle Sam was moving out. Any buyers?

Nobody in the oil business manifested enthusiasm. The public began to sense that something must be wrong. Investigation began. As the evidence unfolded, the military began to admit they had blundered. Among other items the engineers had laid small-diameter pipe incapable of carrying sufficient oil to make commercial operation feasible. If anybody wanted to run the line (and the owners of the wells at Norman showed no inclination to do so), it would be necessary, first, to tear up the existing pipe and start again with a tube of larger bore. The root of the matter, the generals admitted, was that they had made a $434,000,000 mistake. And that was the end of Canol, its pipeline, its winter road south to the Alaska Highway, its landing fields, its camps, its townsite. Canada's Imperial Oil finally took over at what in private business would be called a "distress" figure. Now even the refinery has moved out and has been trucked down the Alaska Highway to a point near Edmonton, where crudes from Canada's new Leduc field are now being processed.

When the Americans pulled out, they left behind equipment to a value of millions of dollars. Bulldozers, jeeps, all manner of heavy engineering gear, cluttered the riverbank across from the wells at Norman. To carry it all south, even if transport could be made available, would have cost more than could be realized in the market for used equipment. Canol became a junkyard monument to military stupidity, carefully alibied by talk of the pressure of events.

The wells and refinery at Norman are peaceful and quiet again. In summer the barges pull alongside and load drums of diesel oil for Eldorado, Yellowknife, and points throughout the country. Wells are capped. Staff is cut. Only tremendous increase in development of the country's resources can bring it back to the bustle it knew in war. Out through the mountains brush grows over the rusted pipe, itself invisible to any but the eye that knows where to find it.

War came and went, and it revolutionized again the business of living in the vast empire of the Mackenzie. With the coming of peace, Canada came for the first time to more vigorous policies to break the country open, realizing its potential value as a source of mineral riches. Peace finds the Mackenzie a country of winter roads and tractor trains, the year-round destination of airliners which land on modern airdromes or on commodious airstrips cut out of the virgin wilderness. In five brief years of war the country was provided with more of the basic necessities of development, particularly improved communications, than had come its way in the preceding two hundred fifty years, or would have been likely to come for decades ahead without the war as its motive power.

Peace found the Mackenzie, in greater degree than ever before, the acknowledged Highway to the Top of the World, the corridor that leads through the new Northwest Passage.

CHAPTER TEN

The Road to the Top

THE MACKENZIE of the end of the first half of the twentieth century is not the river its discoverer found toward the close of the eighteenth. It draws its water from the same sources, they join each other at the same pinpoints on the map, and they follow the same courses to the Arctic Sea, but there the resemblance ends.

During the interval it had been three rivers; first, the river of the Indian, concerned only with his own simple existence; second, the river of the fur traders, concerned principally with the profits of their traffic; third, the river of the prospector, the miner, the frontier buster, and the short-range airplane. The end of a world war in the middle

1940's opened what may be called the era of the Fourth Mackenzie, in which it became, not merely one of the great water systems of a continent, but in every sense of the term a world river.

The same statement could be made in another way. When Alexander Mackenzie found his river, it was a waterway of the flat, or Mercator, map, and it led to nowhere but the icefields of the polar seas. Beyond lay the great emptiness of the Far North. What the men of the air discovered, a century and a half later, was a river on the global map, the curved map, the true map of the world, and they found it in a world they themselves had contracted in size in a ratio of, say, one to one hundred, or more.

When Mackenzie set out in his birch-bark canoe to find his river, forty days and forty nights were consumed in the voyage from Fort Chipewyan to the sea. Today's globe girdler passes through the Mackenzie corridor, across Alaska, and out over the Pacific in the brief span of a few hours.

The great hinterland of Canada's Northwest Territories sits in the dead center of the world's most populous area, which hangs suspended from the Pole in the Northern Hemisphere, where the great density of world population and the major part of its production is found between the 40th and 60th parallels of latitude. We are living, in short, in the age of the New Geography, and of that geography the Mackenzie and the airdromes of the Staging Route are a trunk highway, leading to the top of the world and on beyond to what men still call the Orient, simply because in order to reach it they once had to travel eastward.

At the beginning of World War II, Moscow was a city which, in the mind of the individual North American, lay somewhere east of New York or Montreal, across the Atlantic, and over on the other side of France and Germany. Four years later Moscow had "moved" and lay in the reverse direction. Moscow, in 1944, was reached by flying northwest across Canada, through Edmonton, over the Peace, up the Staging Route, through Alaska, and across Bering Strait into Siberia, a highway over which more than six thousand planes flew to stem the German tide at Stalingrad. China and Japan are no longer places on a map on the far side of thousands of miles of water, but parts of a shrunken world reached by flying the big curve northwest and down the corresponding bend on the far side. Although men still talk, from habit, of Western and Eastern Hemispheres, they are no longer the vital zones. Those were the hemispheres of the Sea Age, when men dug canals through isthmuses or sailed thousands of miles, creeping around the tips of continents to reach their destinations on the other side of the world. Today and tomorrow's hemispheres are the Northern and the Southern, and into the former are packed most of the world's productive people and their means of production. The heartland of the Air Age, in fine, is not in the east or the west, as those terms are generally used, but in the north.

Thus the Mackenzie is the western main road of the Air Age leading out from the world's greatest productive area, the United States of America and, in lesser degree, its neighbor, Canada. But if it is the way out, it is also the way in; and in an uneasy world, which still has not learned that wars must cease as an instrument of policy between groups of people if man himself is to survive, the cold fact that the Mackenzie country is the way *into* North America as well as out of it is a factor which gives men living on that continent pause for urgent thought.

If and when the North American continent should be attacked by a Pacific power again, that attack will be launched through the Mackenzie Valley, or the country that surrounds it if it should come through the air in the form of guided missiles. If it should be an attempt to land armed forces on the ground, then the landing will be made at coastal points in the far northern wilderness, with a view to driving into the flatlands of the Mackenzie country to establish beachheads from which missiles can be launched.

Admittedly points of view on such statements as those made in the preceding paragraph are not held unanimously, but they, or reasonable facsimiles, have been expressed by members of the American, British, and Canadian General Staffs on the highest levels and are matters, therefore, for most serious consideration.

It has not been easy to set such things down on paper in the spring of 1948, when the North American world was in a state bordering on frenzy in respect to its relations with the people living on the other side of the polar highways. What is said here is not written to fit these special circumstances, however, but to make the point: First, that if at any time in the future the North American nations should go to war, no matter with whom, the Far North is this continent's soft underbelly. If North America were to come to mutually satisfactory terms with Russia, the need to defend the arctic would not disappear. It will never disappear so long as aggression remains possible. The threat could come, for example, from a recovered Japan, which is within the realm of foreseeable possibility, in the light of the quaint belief that the Japanese can be democratized and made peace-loving under prototypes of the men who were responsible for Pearl Harbor. Not long ago the United Nations were reported to be investigating a charge that ranking Nazi nuclear scientists, who had escaped to Spain before the debacle, were carrying on their work in Bilbao, with uranium ores made available from the deposits in the Iberian Peninsula. Other German scientists were known to be in the Argentine and what they may have been up to since they arrived there no one on the North American continent seems to know at the time this is written. You may be sure, however, that so long as a remnant of the aggressive mind or the spirit of revenge remains, anywhere in the world, North America will be under potential threat and the arctic and subarctic will be its soft underbelly.

Traditionally, North Americans are not defense-minded. If they are in that mood today, because of the news they read and the broadcasts they hear, removal of the immediate underlying reason undoubtedly would change millions of minds overnight. Even in the circumstances that existed in the spring of 1948 the average North American still refused to regard the Far North as the possible scene of invasion, but, as always, turned his eyes to Europe, to the Middle East, to almost anywhere but the soil of his own continent, always to places on the old, flat map, never to the global charts of the Air Age.

Where, people ask, is an invader going to land in this vast wilderness? The answer is simple. Where did he land before? He landed in the Aleutians, and if the Japanese had sent their major thrust north toward the American coast instead of south, there is no telling what might have happened. As it was, the enemy made effective landings, of which little was heard at the time, either in the continental United States or in Canada, presumably because the governments in Washington and Ottawa did not deem it advisable to let the inland population know what was afoot. Even after the Japanese invasion – and invasion it was – had been successfully repelled, the business was dismissed by the military as if what had happened had been a minor skirmish. It was nothing of the sort. It was a direct threat to continental integrity and it was beaten back because overland communications, which enabled the United States and Canada to throw enough troops and material into the region to sustain a defense and finally to throw the Japanese out, were established in the nick of time. America was extremely fortunate in the Far

North last time. There is no guarantee that such luck would hold good again.

By the end of World War II, Canada and, therefore, the United States, since the Dominion is Uncle Sam's main line of defense, had lagged far behind their arctic neighbor, the Soviet Union, in subzero military know-how. The defense combination to which the job had always been entrusted, geography and climate, had been relegated to oblivion by the long-range airplane, by the rocket, and by atomic fission. What North America needed was a completely new concept and technique of defense. The Soviet Union, the only other arctic power, already had such a concept and the know-how to go with it. North America was starting from scratch.

The machinery was set up in a Joint Defense Board, which is primarily consultative, but which nevertheless moves sharply. The relationship it creates is necessarily delicate, however. Only slightly more than twelve million Canadians occupy a country larger in area than the continental United States. Against any major aggressor they obviously could not defend their own soil. Moreover, any attack launched into the underbelly, though it must go either through or over Canada, would have as its objective the great industrial output of the United States. That is the target anybody who may go to war with the U. S. A. in future is going to attempt to knock out, for that is where World War II was won. Hence the first line of defense of the Republic against attack upon itself is on the soil of another, but friendly, power. And the last people on earth to expect the United States to line its army up south of the 45th parallel of latitude and await the coming of the invader would be the Canadians, if only because by that time they would have been annihilated themselves.

Another factor needs to be noted, however, and that is that the agreement between Canada and the United States is specifically for defense. The arrangement into which the two countries have entered involves either one coming to the help of the other in the event of one of them being attacked from without. It has its roots in a declaration made by the late President Roosevelt in a speech to Canada's Queen's University shortly before the Dominion went to war in 1939, in which the president declared that in the event of attack upon Canada the United States would come to her aid. That promise was confirmed a few days after Canada's declaration of war against the Third Reich in a mutual defense agreement, in which the two nations solemnly pledged themselves to assist each other in the event of attack. What needs to be noted, however, is that neither undertook to go to war *unless its neighbor should be actually attacked*, in proof of which is the fact that Canada was at war with Germany from September, 1939, when the original agreement was made, whereas the United States did not become involved until two years and three months later, and for no reason having directly to do with Canada.

This is said to bring the question of arctic defense into focus. A declaration of war by the United States on any other power does not commit Canada to participation nor require Canada to provide the United States with a jumping-off place for attack until such time as the United States itself is attacked. There has been a great deal of misunderstanding of this on both sides of the border. But it is the essence of the contract, although it is difficult to imagine how Canada could prevent the Army and Air Force of the United States making use of Canadian soil for such purpose if the U. S. were determined to do so, or, to be realistic, withhold from any conflagration in which the United States was a participant. The very fact that Canada has permitted the United States to establish bases in northern Canada *for defense purposes* would obviously make it extremely difficult to persuade the Americans to go home, simply because the condition of attack on North America had not been met. Canada would nevertheless be within her rights under the Agreement should she try to maintain an uneasy peace, and it would be uneasy in the extreme, until the conditions of the contract had been fulfilled by enemy action.

The purpose here, however, is not to debate the merits of the Joint Defense Agreement, but simply to set forth what it involves, before examining the vulnerability of the arctic as a whole and the Mackenzie Valley and its tributary area in particular. Is this vast empty empire defensible?

For answers to such questions it is necessary to rely on the word of the experts, and the experts are not saying much, for obvious reasons. It is clear, however, that the country is not defensible by the deployment of huge armies in a new Maginot Line, dug into arctic ice and bristling with guns to repel an invader trying to force his way up the valley and south to the American border. No military man thinks in such terms, but a great many people do and it is a concept which has no relationship of any kind with the facts.

If the arctic and subarctic are to be defended by forces on the ground, they must be highly mobile forces, capable of being whisked posthaste to the point at which an attack develops. Troops stationed on the coast itself would be used to repel actual invasion from the sea. But nobody believes that is how an attack would come, although it must be provided for. Even attack for purposes of occupation would undoubtedly come by air, via parachute or glider landings, for which the obvious counterattack would itself be from the air. Thus the essence of defense in the northwest is aerial, and it is to provision of superior aerial defenses that the United States and Canada are primarily committed as partners in the north.

The region is also the obvious base for another form of defense. It could be called defense by interception and, again, it is aerial; its purpose is to join issue with aircraft attempting to break through to the industrial centers of the United States for the purpose of laying them waste. In this bracket, too, comes defense, again primarily by air, against the guided missile, a technique developed magnificently by the Royal Air Force, particularly considering the haste that had to be employed, in worsting the attacks on Britain with the robot bomb. In what degree this might be successful in the future, the writer does not profess to know, simply because neither he nor anyone else can tell what tomorrow's aerial weapons of attack will be.

Obviously, then, Alaska and the Canadian Northwest are the first line of North American defense, a defense requiring the utmost mobility, widely deployed, and therefore, in large degree confined to the air. Military activities on the ground, such as Exercise Muskox, which was primarily a test of military vehicles under winter arctic conditions, do not tend to lead the soldiers to any greater use of ground transport than is required by immediate circumstances, excepting in such sections as have been equipped with highways. Cross-country tests have proved painstakingly slow.

But what of the country itself in respect to those things by which man lives, not by which he seeks his own destruction? Will great cities rise in the subarctic north, great industrial centers comparable to those where minerals have been produced in the south (and when you speak from the region of the Mackenzie, "south" means Michigan, or Ontario, or Wisconsin, not Georgia or Tennessee)? The answer must be qualified, but there are clues to the future in events which have happened already. When Gilbert LaBine was on the point of bringing Eldorado to production at Great Bear Lake, he did not establish the full processing plant required at the site of his ore, but only that needed for the first stage of concentration to reduce raw rock, at a rough estimate, 1,000 to 1. To have gone beyond that point would have entailed not merely bringing into the interior everything required in the middle and final stages of refinement to produce either radium or uranium salts or, later, the material that releases U-235. It would have entailed, as well, carrying into the country far greater quantities of materials, ranging from chemicals to foods, to support a doubled or trebled staff. Obviously, then, it was cheaper to carry concentrate from the Arctic Circle to a point in the east, handy to these supplies, than to assemble supplies in the east and carry them to the arctic. Which is precisely what LaBine did. Later, when Canada established its own headquarters for atomic research and production, it placed them in the east, on the free-flowing lines of communication provided by railways and trunk roads.

In the production of gold, however, which down to here remains the principal mineral product of the country, circumstances differ. Gold is most economically refined at the place where it is mined, because, excepting particularly complex ores, the final product, bullion, is obtained without the use of huge quantities of chemicals and with the services of fewer people than are occupied in getting rock out of the ground. All mines brought to the production stage carry out the complete milling and refining process at the minesite, shipping out their gold bricks by air to the Mint at Ottawa.

Because of the high costs of operation at points several hundred miles from the nearest railway, in country without roads, it is impossible to produce minerals profitably from low-grade ores. Hence, the search, speaking generally, has always been for ore bodies rich in mineralization. This leads, of necessity, to the kind of mining the miners themselves would describe as "picking the eyes out of the country." In the area north of Great Slave Lake, where the greatest activity has been and is, however, mineralization tends to be rich, and in some cases extremely rich. Thus, in a country in which potential mining fields are measured in terms of hundreds of square miles, the problem of robbing the richest ground is not likely to occur for decades, if not centuries, to come.

In recent years, particularly since the end of World War II, Canada tends to pursue a much more aggressive policy than was the case in the past in respect to development of its northwestern empire. Throughout the early years of exploration by air, little, if any, encouragement was given to the mine developer, the prospector, or the operator of an airline. Politically-minded men who did not know the north regarded the search for minerals in the arctic wastes as a foolhardy project which could end only in disaster for those so foolish as to squander their capital. That outlook did not change greatly until the war years, when the necessities of defense sent into the country men who had no previous knowledge of the hinterland and who came out extolling its possibilities. Now Canada's government begins to realize that in its frigid back yard lies what may be the richest supply of untapped mineral resources in the world. Canada, therefore, is beginning, at last, to give aid to those who seek those riches, by providing greater facilities for communication and transportation, by utilizing estimates of eminent geologists as to where to look for minerals, and by establishing lower costs for getting the necessities of mining and of life into the country. The empire of the Mackenzie stands on the threshold of the most tremendous development in the history of the country. The result will not be one or more big cities, but many small communities grouped about the mines.

That development will mean the end of the fur trade in the regions where it is still carried on, for when people move in the wild animal life moves on, and with it the trader in fur. The revolution that the fur barons resisted with such doughty, and often ill-judged, vigor has happened. The river of fur has become the river of gold, drawn down from its tributary waters. The river that once flowed to Nowhere has become the Highway to the Top of the World.

Bibliography

Adam, Mercer G., *The Canadian Northwest,* Toronto, 1885.

Back, Sir George, *Narrative of an Arctic Land Expedition to the Mouth of the Great Fish River and along the Shores of the Arctic Ocean in 1833, 1834, and 1835,* London, 1836.

Begg, Alexander, *History of the North-west,* Toronto, 1894.

Bryce, George, *Remarkable History of the Hudson's Bay Company,* London: S. Low, Marston and Co., 1900.

Burpee, L. J., *The Search for the Western Sea,* New York: The Macmillan Company, 1936.

Campbell, Roderick, *Father of St. Kilda,* London, 1901.

Cameron, Agnes D., *The New North,* New York: D. Appleton and Co., 1910.

Camsell, Charles and Malcolm, Wyatt, *The Mackenzie River Basin,* Ottawa: J. de L. Tache, printer, 1919.

Cochrane, Charles N., *David Thompson, the Explorer,* Toronto: The Macmillan Company of Canada, 1924.

Cocking, Matthew, *Journal of Matthew Cocking,* Toronto, 1908.

Crouse, Nellis M., *The Search for the Northwest Passage,* New York: Columbia University Press, 1934.

Cyriax, Richard J., *Sir John Franklin's Last Arctic Expedition,* London: Methuen and Co., 1939.

Douglas, T., *A Narrative of Occurrences in the Indian Countries of North America.*

Dugas, Abbe G., *The Canadian West. . . . ,* Montreal: Librarie Beauchemin, 1905.

Franklin, Sir John, *Narrative of a Journey to the Shores of the Polar Sea in the years 1819-22,* London: J. Murray, 1824.

——, *Narrative of a Second Expedition to the Polar Sea in the years 1825-27,* London: J. Murray, 1828.

HEARNE, SAMUEL, *A Journey from Prince of Wales's Fort in Hudson's Bay to the Northern Ocean,* Toronto: Champlain Society Publication, 1911.

HAWORTH, P. L., *Trailmakers of the Northwest,* New York: Harcourt, Brace & Company, 1921.

INNES, A. J. and ADAMS, HAROLD, *The Fur Trade of Canada,* Toronto: Oxford University Press, Canadian Branch, 1927.

Journals of Samuel Hearne and Philip Turnor, Toronto: Champlain Society Publication, 1934.

KING, RICHARD, *The Franklin Expedition from First to Last,* London, 1855.

LAUT, AGNES, *The Conquest of the Great Northwest,* London: Hodder & Stroughton, 1909.

MACKAY, DOUGLAS, *The Honourable Company; a history of the Hudson's Bay Company,* Indianapolis: The Bobbs-Merrill Company, 1936.

MACKENZIE, SIR ALEXANDER, *Voyages from Montreal* (reprint), New York, 1902.

MIRSKY, JEANNETTE, *To The North!* New York: The Viking Press, 1934.

RAE, JOHN, *Narrative of an Expedition to the Shores of the Arctic Sea, 1846-47,* London, 1850.

RUSSELL, FRANK, *Explorations of the Far North,* Iowa City: The University Press, 1898.

SELKIRK, EARL OF, *A Sketch of the British Fur Trade in North America,* New York: J. Eastburn & Co., 1818.

SIMPSON, ALEXANDER, *Life and Travels of Thomas Simpson, the Arctic Discoverer,* London: R. Bentley, 1845.

SIMPSON, SIR GEORGE, *Narrative of a Journey round the World during the years 1841-42,* Philadelphia, 1847.

SIMPSON, THOMAS, *Narrative of the Discoveries on the North Coast of America . . . during the years 1836-39,* London: R. Bentley, 1843.

Index

Activities, peacetime, 239-40
Adventurers, Gentlemen. *See* Hudson's Bay Company
Aeronautics. *See* Airplanes
Aged, Indian, disposal of, 26-27
Ages, Air and Sea, 243-44
Air route of today, 122
Airplanes, 10-12, 13, 136, 157, 158, 159, 166-208
  airdromes, 13, 224; Northwest Staging Route, 211
  American-made, delivery to Russia, 223
  defense, aerial, 250-51
  difficulties of flying in the Northwest, 221-23
  Eskimos in flight, 176
  fields, landing, 224
  first flight over the North Magnetic Pole, 188-92
  first flight to the delta, 175
  first flight to Great Bear Lake, 181
  fur freighting, 174, 180
  landing fields, 13, 211, 224
  lines: Canadian Airways, 178, 193, 223
    Commercial Airways, 193, 194
    MacKenzie Airways (Brintnell's), 206
Airplanes: lines: Western Canada Airways, 173, 178, 189, 193
    Yukon-Southern Air Transport, 212
  mail contract, first, 193, 194
  mail service, 173, 179
  McKee Trophy awards, 207-08, 223
  mercy flights, 203-07
  mining regions, access to, 183-84
  Northwest Staging Route, 211
  pioneer flight of Brintnell, record of, 176-78
  postal service, 173, 179; first contract, 193, 194
  route, Alsib, and building of 212-21
  trophy, McKee, awards of 207-08, 223
  Western Canada Airways, 173
  Yukon-Southern Air Transport, 212
Akaitcho, 74
Aklavik, 4, 9, 43, 72, 136, 154, 159, 162, 176, 179, 188, 194, 195, 196, 197, 198, 205, 206, 210, 243
Alaska, 132
  air route to, building of, 212-21

Alaska: and boundary dispute, 151
Alaska Military Highway (Alcan), 13, 159, 211, 225-27
Alaska-Siberia plane route (Alsib), 221
Albany (steamship), 52
Albany River, 88, 89
Alberta, 154
Alcan Highway. *See* Alaska Military Highway
Alcoholic liquor for the Indians. *See* Rum
Aleutian Islands, 225; invasion, 246
Alsib (Alaska - Siberia) plane route, 221
American Fur Company, Astor's, 140
Amundsen, Roald, 191
  voyage through the Northwest Passage, 56-57
Arctic Circle, 5, 8, 9, 116, 162, 169
  aviation over, 175
Arctic exploration. *See* Northwest Passage
"Arctic exploration, father of" (Sir John Barrow), 57, 60, 82
Arctic Red River, 9, 43, 119, 132, 144, 146, 156, 162, 195
  burning of post, 194
Arctic regions, British interest in, 59-60, 62, 72, 81-82
Arctic Sound, 64
Astor company, the American Fur Company, 140
Athabaska (steamboat), 154, 159
Athabaska country, 88, 102
Athabaska Lake, 5, 11, 16, 29, 33, 48, 59, 91, 99, 114, 140, 158, 166, 236
Athabaska Lake-Chipewyan region, 88, 102
Athabaska River, 5, 10, 11, 16, 119, 153, 169, 210, 224
Athabaskan Indians, 16, 90
Atom bomb, 4, 6, 155, 160, 172, 202, 230-31
  ore for, 4, 6
Atomic research, 230-31
Aurora Borealis, 222
Aviation. *See* Airplanes

Back, Commander Sir George, 63, 64, 66, 67, 70, 73, 178
  expedition in search of Ross, 73-75
  log, 80-81
  river named for. *See* Back's (Great Fish) River
Back Bay, 192
Back's (Great Fish) River, 73-76, 78, 84, 85
  exploration of, 74-76
Baffin, William, and Baffin Bay, 58, 61
Bannerman, Doctor, 182, 183
Barbour and Knight, expedition, 52

Barren Lands, 64, 207
Barrier River, 197
Barrow, Sir John, "Father of Arctic exploration," 57, 60, 82
Barrow Point, 71-72, 76, 77
Barrow Strait, 83
Basin of the river, 87, 88
Battle of Seven Oaks, 108
Bay Company. *See* Hudson's Bay Company
Baychimo (steamboat), 190
Bear River, 70, 132, 200
Beaufort, Fort, 9
Beaufort Sea, 7, 9, 32, 45, 154
Beaver Indians, 16, 48, 117
Beechey, Captain Frederick, 67, 71, 76, 82
Beechey Island, 84
Bell, John, 131
Bend, Great (Camsell), 8
Bering, Vitus, 59
Bering Strait, 60, 67, 243
Berry, Matt, 170
  mercy flight, 204-07
  rescue of Con Farrell, 207
Big Island, 38, 131, 132, 153, 160, 161
Bishop Lake, 205
Black, Samuel, 115, 117
Blackfoot Indians, 138
Bloody Falls, 54
Blossom (ship), 67
Bluefish River, 40
Boat, York. *See* York boat
Boats, river. *See* Canoes; Steamboats; York boats; names of vessels
Boland's Camp, 181
Bomb, atomic, 46, 155, 160, 172, 202, 230-31
  ore for, 4, 6
Boothia Felix Peninsula, 80, 191
Bowen, Jack, 197
Brabant, Port (Tuktuk), 154, 155, 162
Brintnell, Leigh, 152, 166, 171, 173, 182, 183, 189, 201
  airline, Mackenzie Airways, 206
  pioneer flight record, 176-78, 180
British interest in the Arctic, 59-60, 62, 72, 81-82
British North America Act, 147
Buchan, Captain David, 60, 62
Buchanan (aviator), 189-93
Builders of empire, new, 166-208
Burnside Harbor, 205
Burwash, Major Lachlin, 189-92, 201

CAT-train, 13
Cambridge Bay, 191, 204
Camera survey of the coast, 192
Cameron Bay, 199
Campbell, Colin, 140
Campbell, Robert, 139-40
Camsell (Great) Bend, 8
Canada, and the Hudson's Bay Company. *See* Free country

Canada: British North America Act, 147
  Confederation, 146-47; and breaking of the Hudson's Bay Company monopoly, 141-42
  defense agreement with the United States, 247-51
  development of, 253-54
Canadian Airways, 178, 193, 223
Canadian Mounted Police, 7, 195-98
Cannibalism, on Franklin expedition, 66
Canoes, 24, 86, 163
Canol (oil pipeline), 4, 6, 8, 9, 12, 13, 40, 159, 236-39
  blunder, 236, 238-39
  cost, 9
  failure, 13
  headquarters, 224
Cape Adelaide, 191
Cape Bathurst, 71
Cape Crozier, 191
Cape Hardy, 191
Cape Krusenstern, 191
Carcross, 226
Carrying places. *See* Portages
Castor (vessel), 76
Castor and Pollux Bay, 78
Channels, 156
Chawchinahaw, 53
Chipewyan, Fort, 18, 29, 32, 34, 43, 48, 51, 63, 67, 70, 114, 115 117, 119, 120, 135, 167, 168, 169, 179, 242
Chipewyan-Lake Athabaska region, 88, 102
Chipewyan Indians, 16, 19, 23, 53
  enemies, hereditary, of the Eskimos, and massacre of Eskimos, 54-55
Churchill, 67, 79
Churchill River, 89
Circle, Arctic, 5, 8, 9, 116, 162, 169
  aviation over, 175
Cities of the future, 251-52
Clair, Lake, 5, 16
Clearwater River, 119, 120, 167
Climate, 133-35
Closing of the river by ice, 7, 35
  going out of the ice, 153, 160
Coast, camera survey, 192
Coast line survey, 71-72, 73, 76
  by camera, 192
Cobalt bloom, 183
Collinson, Captain Richard, 83
Colville, Neil, 114
Coltman, Colonel, 110
Commercial Airways, 193, 194
Committee (Selkirk) Bay, 79, 80
Communications, 121, 134
Confederation. *See* Canada
Confidence, Fort, 77, 78, 181
Conga-ca-tha-wha-chaga River, 54
Conservation policy, Hudson's Bay Company, 126
Cook, Captain James, 59
Copper, 51, 55
Copper Indians, 54

Coppermine River, 30, 50, 52, 59, 62, 63, 64, 70, 71, 73, 75, 77, 78, 85, 89, 190, 191, 192, 205, 236
discovery, 54
Cording, Matthew, 89
Coronation Gulf, 30, 50, 63, 77, 85, 205, 236
Cost of travel, 179
Country, free, 141-42, 143-65
Cowey Lake, 54
Cree Indians, 53
Croker Mountains, blunder of Ross, 61, 72
Cumberland House, Hudson's Bay Company post, 89
Currents, strong, 39
Curry, Thomas, 89
Cyriax, Richard J., on the Franklin expedition, 57-58

Darnley Bay, 205
Davis, John, and Davis Strait, 57-58, 60
Dawson Creek, 212, 226
Daylight, long, 38
Dease, P. W., 70, 73, 137, 171, 178
Dease Arm, 77, 181
Dease Lake, 219
Dease River, 220
Dease, Simpson, and Rae expeditions, 75-80
Deerpass Bay, 181, 182
Defense of the territory, 12-13
Defense, aerial, 250-51
Defense agreement, United States and Canada, 247-51
Defense, continental, role of the Northwest, 209-40
Defense Board, Joint, 247
Defense-mindedness, 245-46, 247
Defensibility of the Mackenzie Valley, 249-50
Delta of the river, 9-10, 43, 119, 136, 154
first flight to, 175
Dennis, Colonel, 151
Dickins, Punch, 11, 157, 171, 173-76, 181-84, 193, 203
winner of the McKee Trophy, 208
Discoverer of the river, Sir Alexander Mackenzie. *See* Mackenzie, Sir Alexander
Discovery (ship), 52
Distributor III (steamboat), 154, 155, 159, 160, 161
last voyage, 161-62
Dog-team service, 134
Dogribs (Indians), 8, 17, 18, 19, 22-23, 24, 25, 40, 117, 156
Dolphin, 70
Dominion of Canada. *See* Canada
Douglas, Thomas. *See* Selkirk, Earl of
Douglas, Fort, 107-09
Dubawnt Lake, 53

Edmonton, city and fort, 137, 153, 162, 168, 169, 170, 173, 198, 202, 206, 210, 212, 222,

Edmonton: city and fort (*cont'd*): 223, 236, 238, 243
  Great House, 137-38
"Edmonton, U. S. A.," 226
Eldorado radium-uranium mine, 6, 170, 180, 185-88, 201, 202, 206, 239, 251-52
  production of uranium, 227-36
Embarras, 224
Empire builders, new, 166-208
"End of steel." *See* Railroad
English Chief (Indian), companion of Mackenzie, 30, 39-40, 43, 44, 47, 48
Enterprise, Fort, 63, 64, 65, 66
Erebus and *Terror* (ships), 80, 81, 84
Erik the Red, voyage of, 57
Eskimos, 16, 17, 188
  airplane flights, 176
  and the Indians, 27
  enemies, hereditary, of the Chipewyans; and massacre of Eskimos, 54-55
  feared by the Indians, 41, 43, 44, 55
  kindness to Back, 75
  trade with, 132
Ethnography, Indian, 16-18, 19-28
Exercise Muskox, 251
Exploration, Arctic. *See* Northwest Passage
"Exploration, Arctic, father of." *See* Barrow, Sir John
Exploration, Arctic, British interest in, 59-60, 62, 72, 81-82
Explorers, early, 30

Fairbanks, Alaska, 13, 223, 226
Falaize, Monsignor, rescue by air, 204-07
Family Compact, 111
Farrell, Con, rescue by air, 207
"Father of Arctic Exploration." *See* Barrow, Sir John
Fishing and hunting, 135
  Indian, 21-22
Fission, atomic, 230-31
Fitzgerald, 5, 34, 154, 200
Flights. *See* Airplanes
Flow of river, to the north, 6
Foggy Island, 72, 76
Food at Hudson's Bay Company posts, 135-36, 143
Forsyth, Richardson and Company (XY Company), 95, 97
Forts: Beaufort, 9
  Chipewyan, 18, 29, 32, 34, 48, 51, 63, 67, 70, 114, 115, 119, 120, 135, 167, 168, 169, 179, 242
  Confidence, 77, 78, 181
  Douglas, 107-09
  Edmonton. *See* Edmonton
  Enterprise, 63, 64, 65, 66
  Franklin, 8, 70, 71
  Garry, 146, 148, 150, 151
  Gibraltar, 105, 106, 107

Forts: Good Hope, 9, 40, 120, 122, 131, 136, 162, 175, 194
 Halkett, 132
 Liard, 132
 McMurray, 119, 153, 154, 167, 168, 169, 173, 175, 179, 181, 188, 193, 194, 196, 198, 200, 206, 210
  landing field, 224
 McPherson, 119, 120, 131, 132, 136
 Nelson, 212, 213, 214, 220, 226
  building of trail to, 211-19
 Norman (Norman Wells), 4, 8, 12, 70, 120, 122, 131, 136, 154, 159, 160, 162, 172, 175, 181, 184, 189, 194, 199, 206
  landing field, 224
  oil at, 237, 238, 239
  site of, 132
 Prince of Wales, 52, 53, 55
 Providence, 8, 70, 132, 156, 161, 179
  landing field, 224
 Rae, 167, 175, 184
 Reliance, 74
 Resolution, 7, 11, 122, 157, 167, 173, 174, 179, 203, 206
  landing field, 224
 St. John, 212, 215, 220, 226
 Selkirk, 144
 Simpson, 8, 39, 70, 120, 121-22, 131, 132, 133-34, 162, 172, 174, 194
 Smith, 5, 7, 34, 154, 156, 159, 161, 162, 167, 170, 174, 200,
Forts: Smith (*cont'd*): 206
  landing field, 224
 Vermilion, 172
 Wedderburn, 114
 William, 109
 Yukon, 131, 140-41
Franklin, Sir John, 55, 56, 57, 178
 advised by Mackenzie, 63
 cannibalism of member of expedition, 166
 death, 84
 descent of the Mackenzie, 67
 expeditions, and loss, 51, 58-85
 first expedition, 62-67; hardships, 64-67
 knighted, 72
 search for, 83
 search for relics, 192
 second expedition, 67-71
 third and last, fatal, expedition, 72, 80-85; relics shown by Eskimos, 83
 tomb, search for, 189
 with the Buchan expedition, 60
Franklin, Fort, 8, 70, 71
Franklin, Lady, 83, 84
Free country, 141-42, 143-65
 and the Hudson's Bay Company, 129-30
Freight rates, 158, 200
Freight routes, 153
Freight transportation, 251-52
Freighters, LaBine's, 200
French-Canadians in the fur trade, 93-94
Frenchmen in the fur trade, 87

Frobisher, Sir Martin, 58
Frobisher and McTavish, 90
Frontier country, 3, 4, 13-14
Frozen Strait, 81
Fullerton, Elmer, 172
Fur trade, 6, 8, 9, 10, 12, 30-31, 86-142
  airplanes, freighting by, 174, 180
  Battle of Seven Oaks, 108
  bloodshed in, 91
  boatmen's food and equipment, 94
  business era, beginning of, 116-17
  competition, renewed, 152
  conservation policy, Hudson's Bay Company, 126
  decline, 87
  end in sight, 254
  French-Canadians in, 93-94
  Frenchmen in, 87
  golden years, 113-42
  independent, 88, 89, 90
  Indians in, 22, 23; trouble with, 115-16
  metals in exchange for furs, 23
  monopoly. *See* Hudson's Bay Company
  Montreal men, 90
  output, Hudson's Bay Company, 130-31
  periods, three, 86-87
  rivalry, 89, 91-112
  Selkirk affair, 102-12
  varieties of furs, 120
Fur trade: war in, 95-112, 114-15; case of D. Rousseau, 97; companies', 95-112
  See also Forsyth, Richardson and Company (XY Company); Hudson's Bay Company; North West Company; McTavish, Simon
Fury (ship), 61
Fury and Hecla Strait, 61, 78, 80, 81

Garry, Fort, 146, 148, 150, 151
Gateshead Island, 83
Geiger counter, 234, 235
Gentleman Adventurers. *See* Hudson's Bay Company
George VI, 209
Germans, relations with, 245
Gibraltar, Fort, 105, 106, 107
Gilbert, Sir Humphrey, 57
Gilbert, Walter, 171, 173, 188-93
  winner of the McKee Trophy, 208
Gilmour (flight engineer), 189-93
Gjoa (ship), 56, 191
Gjoa Haven, 191, 192
Gold, 4, 6, 11, 155, 160, 167, 168-69, 201-03, 252
  rush at Yellowknife, 175
Goldfields, 167-68, 169, 201
Good Hope, Fort, 9, 40, 120, 122, 131, 132, 136, 162, 175, 194
Gorman, George, 172

Grahame (steamboat), 154, 156
Grand Prairie, 212, 220, 226
Grande Portage, 90, 95, 97, 116
Grant, Cuthbert, Sr., 35
Grant, Cuthbert, Jr., 108
Great Bear Lake, 5, 6, 8, 11, 12, 13, 70, 77, 85, 155, 157, 158, 159, 163, 169, 171, 180, 184, 188, 189, 191, 192, 193, 199, 202, 204, 206, 227, 231, 232, 233, 251-52
  first flight to, 181
  mining boom, 199, 201
  region, source of pitchblende, 186
  storms, 200
Great Bear River, 8, 40, 181
Great Bend (Camsell), 8
Great Britain, interest in Arctic exploration, 59-60, 62, 72, 81-82
Great Falls, Montana, 223
Great Fish (Back's) River, 73-76, 78, 84, 85
  exploration of, 76-78
Great House at Edmonton, 137-38
Great Slave Lake, 3, 4, 5, 7, 11, 17, 18, 23, 30, 34, 35, 47, 63, 67, 70, 73-74, 85, 99, 108, 120, 122, 132, 133, 153, 155, 157, 160, 167, 172, 175, 184, 201, 204, 253
  crossing by Hearne, 55
  outlet, 131
Great Slave Lake: railroad across frozen lake, projected, 203
  storm on, 47-48
Greenland, 60
  discovery of, 56
Gregory, John, 33
Gregory and McLeod, 90

Half-breeds (Metis), 148
Halket, Fort, 132
Hardships of first Franklin expedition, 64-67
Hare Indians, 3, 17, 41, 117, 156
Haworth, P. L., *Trailmakers of the Northwest*, 80
Hay River, 38, 160, 161, 172, 179, 202, 203
Hearne, Samuel, 88, 89, 137, 171
  crossing of Great Slave Lake, 55
  explorations, 30, 50, 51, 52-56, 59, 62, 63, 72
Hecla (ship), 61, 62
Hecla Strait, 61
Hepburn, John, 63, 65, 66
Herschel Island, 188
Hersey, Sergeant, H. F., 197
Hervey, James, *Meditations*, 139
Highway, Alaska-Canada, 159, 225-27
Highway *into* North America, 244-45
Highway to the top of the world, 240, 241-54
Hood, Robert, 63, 65

Hood, Robert: murder of, 66
Hood River, 64
Hornaday River, 204
House, Great, at Edmonton, 137-38
Howe, the Honorable C. D., 227-30, 231, 232
Hudson, Hendrik, 49, 58
Hudson Bay, 7, 33, 52, 59, 67, 87, 88, 89, 105, 114, 119
  route for fur trade, 116
Hudson Strait, 50, 60, 61, 81, 82
Hudson's Bay Company, 30, 50, 51, 52, 58, 70, 73, 74, 78, 79, 83
  amalgamation with the North West Company, 86, 111-12
  advantages; control from London, 99
  and Canada, 129-30. *See also* Free country
  and its rivals, 32-34, 91-92, 98-99
  charter, 87, 146
  competition from the North West Company, 98-99
  conservation policy, 126
  Cumberland House, 89
  earnings, 129
  forts. *See* Forts; Hudson's Bay Company: Posts
  Indian policy, 125
  men of, 137-41
  methods and policy, 124-26
Hudson's Bay Company: monopoly, 124-25, 130-31, 144-45; breaking up of, after Canadian Confederation, 141-42, 143-65
  movements against rule. *See* Free country
  organization under Simpson, 128
  output of furs, 130-31
  posts, 131-37; life in, 132-37
  rivalry with the North West Company, 91-92
  Selkirk control, 104
  settlement with the Government; terms, 147
  Transfer Act, 147
    See also Fur trade
Hunting and fishing, 135
  Indian, 21-22

Ice, closing of the river by, 7, 35
  going out of, 153, 160
Iceland, discovery of, 56
Icy Cape, 59, 67, 70, 73, 76
Immigration, Scottish, to Canada, 103-12
Imperial Oil Company, 238
Indians, afraid of the Eskimos, 41, 43, 44, 55
  aged, disposal of, 26-27
  Athabaskan, 16, 90
  Beavers, 16, 48, 117
  Blackfoot, 138
  Chipewyans, 16, 19, 23, 53
  contact with the whites, 24-25

Indians: Copper, 54
  Crees, 53
  Dogribs, 8, 17, 18, 19, 22-23, 24, 25, 40, 117, 156
  Eskimos, relations with, 27
  ethnography, 16-18, 19-28
  fear of the white men, 17, 18, 42
  fishing and hunting, 21-22
  fur company employees, 92
  half-breeds (Metis), 148
  Hares, 3, 17, 41, 119, 156
  hunting and fishing, 21-22
  petitions in their interest, 145-46
  policy of the Hudson's Bay Company, 125
  "Quarrelers," 42
  queen, 139
  rum question. *See* Rum
  Simpson's opinion of the Indians, 118
  Slaveys, 3, 8, 16, 17, 156
  women, status of, 25-26
Insurrection, Riel's, 148-53
Invasion by the Japanese, 246
Investigator (ship), 56, 83
Isbister petition, 145-46
Islands: Aleutians, 225; invasion by Japanese, 246
  Beechey, 84
  Big, 38, 131, 153, 160, 161
  Foggy, 72, 76
  Gateshead, 83
  Herschel, 188
Islands: King William's Land, 72, 73, 78, 83, 84-85, 88, 189, 190, 191, 192
  Matty, 189, 191
  Melville, 61, 62
  northern, 76
  Reid, 204
  Victoria, 77, 191
  Whale, 45

James Bay, 87, 88
Japan, invasion by, 246
  relations with, 245
Joachimsthal, source of pitchblende, 186
Johnson, Albert, "Mad Trapper," story of, 195-98
Joint Defense Board, 247

Kamloops, 220
Kendall, E. N., 67
Kent, Duke of, 97
King George VI, 209
King, Dr. Richard, 73
King William's Land (Island), 72, 73, 78, 83, 84-85, 88, 189, 190, 191, 192
Knight (flight engineer), 189-93
Knight and Barbour expedition, 52

LaBine, Gilbert, 5, 11, 157, 158, 163, 169-70, 171, 180-81, 182, 183, 184, 185-88, 198-201, 227-36, 251-52

LaBine, Gilbert: discovery of pitchblende, 185-88
  Eldorado mine. *See* Eldorado
LaBine Point, 184
Lady Franklin Point, 191
Lakes: Athabaska, 5, 11, 16, 29, 33, 34, 48, 59, 91, 99, 114, 140, 158, 166, 236
  Bishop, 205
  Clair, 5, 16
  Cowey, 54
  Dease, 219
  Dubawnt, 53
  Great Bear, 5, 6, 8, 11, 12, 13, 70, 77, 85, 155, 157, 158, 159, 163, 169, 171, 180, 181, 182, 183, 184, 186, 188, 189, 191, 192, 193, 199, 202, 204, 206, 227, 231, 232, 233, 251-52
    mining boom, 199, 201
    storms, 200
  Great Slave, 3, 4, 5, 11, 17, 18, 23, 30, 34, 35, 47, 63, 67, 70, 73-74, 85, 99, 108, 120, 122, 132, 133, 153, 155, 157, 160, 167, 172, 175, 184, 201, 204, 253
    crossing by Hearne, 55
    outlet, 131
    railroad across frozen lake, projected, 203
    storm on, 47-48
  Lesser Slave, 5
  Mill, 224
Lakes: Pishew, 54
  Red, 172
  Superior, 90, 96, 99
  Watson, 212, 219, 220, 221, 222, 226
  Williams, 220
  Winnipeg, 107
  Winter, 63
La Loche Portage, 114, 119
Lancaster Sound, 58, 60, 61, 82
Length of the river, 5, 7
Le Roux, Laurent, companion of Mackenzie, 30, 35, 48
Lesser Slave Lake, 5
Letty Harbor, 205
Liard, Fort, 132
Liard River, 8, 39, 47, 120, 131, 132, 139, 162, 220, 226
Lief Erickson, voyage of, 57
Liquor for the Indians. *See* Rum
Lord Mayor's Bay, 80
Lower Ramparts, 43, 119
Lyon, Captain, 81

MacDonnel, Captain, 105, 106
Mackenzie, Sir Alexander, and the Indians, 17, 18, 24
  backing by Montreal men, 31
  consulted by Franklin, 63
  discovery of the river, 29-48
  fortune, search for, and success, 31-32
  interest in furs, 30-31
  Journal, 15-16, 70, 97, 176
  knighted, 97

Mackenzie, Sir Alexander: on Hudson's Bay Company rivalry, 91
pioneer, 38
prestige in the Northwest, 102
second voyage, 94
service with the North West Company, 90
writings, 15-16, 70, 97, 176
Mackenzie, Dr. C. J., 230
Mackenzie, Roderick, cousin of Sir Alexander, 34
Mackenzie Airways, Brintnell's, 206
Mackenzie Range, 8
Mackenzie River (steamboat), 154, 159, 160, 162
MacLaren, Gil, 206
"Mad Trapper," story of, 195-98
Magnetic Pole, North, discovery of, 72
first flight over, 188-92
Mail contract, air, first, 193, 194
Mail service by plane, 173, 179
Manitoba, 152, 180
Massacre of Eskimos by Chipewyans, 54-55
Matonabee, 53-54
Matty Island, 189, 191
May, Capt. W. R. ("Wop"), 157, 171, 173, 193-98, 222
part in killing of the "Mad Trapper," 195-98
winner of the McKee Trophy, 208
McClintock, Captain, 84
McClure, Capt. Robert, completion of the Northwest Passage, 56, 83
McConachie, Grant, 212
McDougall, the Honorable William, 149-50, 151-52
McKee Trophy (aviation), 207-08, 223
McLeod, John, 107
McMurray, Fort, 119, 153, 154, 167, 168, 169, 173, 175, 179, 181, 188, 193, 194, 196, 198, 200, 206, 210
McPherson, Murdoch, 140
McPherson, Fort, 119, 120, 131, 132, 136
McTavish, Simon, 94-95
McTavish Arm, 183, 184
Meditations, by James Hervey, 139
Melville Island, 61, 62
Melville Peninsula, 61
Mercy flights. *See* Airplanes
Metals traded for furs, 23
Metis (half-breeds), 148
Miller, Dr. W. G., 186
Mills, Lake, 224
Minerals. *See* Mining; names of minerals
Mining, 158, 180
airplanes, access by, 183-84
at Great Bear Lake, boom, 199, 201, 233
ores, rich, search for, 252-53
progress of, in the Mackenzie region, 170

Mining: prospecting, scientific, 233-36
See also names of minerals
Mirsky, Jeannette, on the Franklin expedition, 64, 65
Missioners, rescue by plane, 204-07
Mississippi River and the Mackenzie, comparison, 156
Monopoly, fur trade. *See* Hudson's Bay Company
Montreal backers of Mackenzie, 31-34, 90
Moscow, 243
Mountains, Croker, blunder of Ross, 61, 72
  Mackenzie Range, 8
  Stoney, 54
Mounted police, 7, 195-98
Mouth of the river at Fort Beaufort, 9-10
Murdoff, D. K., 179
Murphy, Dick, 234
Murray, Alexander Hunter, 140-41
Muskeg, the, 8, 214, 217
Muskox, Exercise, 251
Muskwa River, 218

Names of the river, 5, 7, 10
National Research Council, 230
Navigability of the river, 5, 7, 34, 38, 156
  season, 35
Nelson Airport, 219
Nelson, Fort, 212, 213, 214, 220, 226
  building of the trail to, 211-19
Nelson River, 89
Newfoundland, 57
Newspaper, *Nor'Wester*, 146
Nig Creek, 214, 215
Night, short, 38
Norman, Fort, and Wells, 9, 12, 40, 70, 120, 122, 131, 136, 154, 159, 160, 162, 172, 175, 181, 189, 194, 199, 206
  landing field, 224
  oil at, 237, 238, 239
  site of, 132
North Magnetic Pole, discovery of, 72
  first flight over, 188-92
North West Company, 30
  advantages over Hudson's Bay Company, 99
  amalgamation with Hudson's Bay Company, 86, 111-12
  amalgamation with XY Company, 98
  competition with Hudson's Bay Company, 98-99
  expansion, 98
  founding of, 33-34, 90
  Mackenzie's service with, 90
  partner, senior, S. McTavish, 94
  rivalry with XY Company, 91-92, 99-102
Northern Lights, 222

Northern Traders Company, 179
Northmen, search for the Northwest Passage, 56, 57
Northwest Passage, completion by McClure, 56
  crossing on ice, by McClure, 83
  search for, 32, 34, 39, 45, 47, 49-85
  voyage through, by Amundsen, 55-56
Northwest Staging Route, 13, 211, 243
Nor'Wester, newspaper, 146
Norton, Moses, 52, 54
Norton, Richard, 52
Norway House, 107, 114, 115
Nyman, Gus, 168-69
Nyman Bay, 169

Oaks, H. A. ("Doc"), 172-73
Oil, finding of, 155
Oil pipeline (Canol), 4, 6, 8, 9, 12, 13, 159, 236-39
  blunder, 236, 238-39
  cost, 9
  headquarters, 224
Oilfields, 4, 40
Open season on the river, 7
Ores. *See* Mining; names of minerals
Our Lady of Lourdes (schooner), 205

Pacific Ocean, route to, 31
Parmenter, Lew, 174
Parry, Edmund William, 60, 61, 67
Parry Archipelago, 61
Peace River, 5, 10, 16, 34, 115, 172, 203, 210, 212, 226, 243
Peace River (town), 172
Peace River country, 5
Peacetime activities, 239-40
Peel's River, 131
Pelly River, 140
Petroleum, finding of, 155
  pipeline (Canol), 4, 6, 8, 9, 12, 13, 40, 159, 236-39; blunder, 236, 238-39; cost, 9; headquarters, 224
Photographic survey of the coast, 192
Pierre's House, 132
Pine-to-Conroy section of the Alaska Highway, 216
Pishew Lake, 54
Pitchblende, 6, 9, 11, 163, 171, 183, 185, 194, 198-201
  production by LaBine mine, 227-36
  rush to scene of discovery, 199
  search for new sources, 233-36
  source, Great Bear Lake region, 186
Planes. *See* Airplanes
Point Barrow, 71-72, 76, 77
Point Turnagain, 63, 70, 72, 78, 80, 81
Pole, North Magnetic, discovery of, 72
  first flight over, 188-92

Police, Canadian Mounted, 7, 195-98
Pollux (ship), 76
Pond, Peter, 33
Population, growth of, 144-45
Port Brabant (Tuktuk), 154, 155, 162
Port Hope, Ontario, 170
Port Radium, 8
Portages, 5, 34, 94, 95, 97, 114, 116, 119, 120, 154, 200
Postal contract, air, first, 193, 194
Postal service by planes, 173, 179
Posts, trading. *See* Hudson's Bay Company
Prince Edward Island, 104
Prince of Wales Fort, 52, 53, 55
Prince Regent Inlet, 61
Prince Rupert, 220
Princess Victoria Cape, 75
Progress in the mining region, 170
Prospectors. *See* Mining
Providence, Fort, 8, 12, 70, 132, 156, 161, 179
  landing field, 224

"Quarrelers" (Indians), 42
Queen, Indian, 139
Queen Maud Gulf, 191

Radium, 5, 6, 8, 157, 158, 160, 169-70, 171, 183, 199, 201
  search for new sources, 233-36
Radium King (steamboat), 200
Radium Queen (steamboat), 200
Rae, Dr. Sir John, 73, 83, 137, 171, 178
  expedition, 78-80
Rae, Fort, 167, 175, 184
Rae Isthmus, 79
Rae, Simpson, and Dease expeditions, 75-80
Rae Strait, 19
Railroad, all-year, projected, 203
  development of, 152-53
  "end of steel," 5, 34, 119, 153, 168, 170, 175, 181, 188, 194, 210, 212
Rampart House, 132
Rapids, 8, 34, 40, 46, 200
  See also Portages
Rat River, 195
"Rebellion," Riel's, 148-53
Red Lake, 172
Red River. *See* Rivers – Arctic Red River
Red River Valley Settlement, 78, 103-12, 148, 150, 151
  "Rebellion." *See* Riel, Louis
Regent's Inlet, 81
Reid Island, 204
Reliance, Fort, 74
Religions, clash of, 148-52
Repulse Bay, 79, 81
Research, atomic, 230-31
Resolution, Fort, 7, 11, 122, 157, 167, 173, 174, 179, 203, 206
  landing field, 224
Return Reef, 76

Richardson, Dr. John, 63, 64, 65, 66, 67, 71
Richardson, James A., 172-73, 178, 180, 187, 193, 199
Riel, Louis, and the Riel "Rebellion," 148-53
 second insurrection, and execution, 152
Rivers: Albany, 88, 89
 Arctic Red, 119, 144, 146, 156, 162, 195; burning of post, 194
 Athabaska, 10, 11, 16, 119, 153, 169, 210, 224
 Back's (Great Fish), 73-76, 78, 84, 85; exploration of, 74-76
 Barrier, 197
 Bear, 70, 132, 199, 200
 Bluefish, 40
 Churchill, 89
 Clearwater, 119, 120, 167
 Conga-ca-tha-wha-chaga, 54
 Coppermine, 30, 50, 52, 59, 62, 63, 64, 70, 71, 73, 75, 77, 85, 89, 190, 191, 192, 205, 236; discovery of, 54
 Dease, 220
 Eagle, 197
 Great Bear, 8, 40, 181
 Great Fish (Back's), 73, 78, 84, 85; exploration of, 74-76
 Hood, 64
 Hornaday, 204
Rivers: Liard, 8, 39, 47, 120, 131, 132, 138, 162, 220, 226
 Muskwa, 218
 Nelson, 89
 Peace, 5, 10, 16, 34, 115, 172, 203, 210, 212, 226, 243
 Peel's, 131
 Pelly, 140
 Rat, 195
 Red. *See* Rivers – Arctic Red
 Saskatchewan, 88, 89, 98, 114
 Sikanni, 213, 214, 216, 217
 Slave, 5, 6, 11, 16, 34, 35, 48, 99, 122, 154, 158, 161, 184
 Stikine, 219
 Thlew-ee-choh. *See* Rivers: Back's
 tributary, 5, 8, 38, 47, 117
 York, 107
 Yukon, 46, 122, 140, 210; discoverer of, 131
Road *into* North America, 244-45
Road to the top of the world, 240, 241-54
Rocks. *See* Mining
Roosevelt, President Franklin D., on defense agreement with Canada, 248
Ross, Sir James, 81
Ross, Sir John, 33-34, 60, 62, 80
 death of, 91
 expedition, 72-74
Rousseau, Dominique, driven out of the fur trade, 97

Route *into* North America, 244-45
Route of the river, 6-10
Route to the Pacific, 31
Route to the top of the world, 240, 241-54
Routes, transportation, 119-21, 153
Rowand, John, 137-39
Royal Geographic Society, 82
Rum for the natives, 19, 90, 92, 97, 122-24, 126
Rupert, Prince, 32
Rupert's Land, 33, 103, 146, 147
Russia, relations with, 244-45
Russians in the Northwest, 140-41

St. John, Fort, 212, 215, 220, 226
St. Paul, E. C., 184, 185
Sans Sault Rapids, 40, 46
Saskatchewan River, 88, 89, 98, 114
Scottish immigration into Canada, 103-12
Scott, Thomas, shooting of, 151
Sea Age and Air Age, 243-44
Selkirk, Earl of (Thomas Douglas) and the Selkirk expedition, 102-12
  death of Selkirk, 111
Selkirk colonists, 144
Selkirk (Committee) Bay, 79, 80
Semple, Governor, 107, 108
Seven Oaks, Battle of, 108
Sherbrooke, Sir John, 110
Siberia, 243
Siers, Tommy, winning of the McKee Trophy, 223
Sikanni River, 213, 214, 216, 217
Silver, 6
Simpson, Sir George, 73, 114-42, 171, 178
  business policy, 143
  death, 78
  on the Indians, 118
Simpson, Thomas, 137
Simpson (town), 203
Simpson, Dease, and Rae expeditions, 75-80
Simpson, Fort, 8, 12, 39, 70, 120, 121-22, 131, 132, 133-34, 136, 162, 172, 174, 194
Sir James Ross Strait, 192
Slave River, 5, 6, 7, 11, 16, 34, 35, 48, 99, 122, 154, 158, 161, 184
Slaveys (Indians), 3, 8, 16, 117, 156
Smallpox, 90
Smith, Donald (Lord Strathcona), 150
Smith, Fort, 7, 34, 154, 156, 159, 160, 161, 162, 167, 170, 174, 200
  landing field, 224
Smithers, 220
Soviet Russia, relations with, 244-45
Spirits for the Indians. *See* Rum
Spitsbergen, 60, 62
Staging Route, Northwest, 211, 243

Steamboating on the river, 87, 153-63
freighters, LaBine's, 200
See also names of vessels
"Steel, end of." *See* Railroad
Steinbruck, John, companion of Mackenzie, 29
Stikine River, 219
Stoney Mountains, 54
Storms, on Great Bear Lake, 200
on Great Slave Lake, 47-48
Strachan, Doctor, 111
Strathcona, Lord (Donald Smith), 150
Summer traffic, 156
Superior, Lake, 90, 96, 99
Survey of the coast by camera, 192
Survey of the river, 7

Telegraph Creek, 219
Teslin, 226
Terror and *Erebus* (ships), 80, 81, 84
Thlew-ee-choh River. *See* Rivers: Back's (Great Fish)
Time required for trading, 141
Top of the world, route to, 240, 241-54
Trading posts, 9
See also Hudson's Bay Company
Traffic, summer, 156
Trail, Nelson, building of, 211-19
Trailmakers of the Northwest, by Haworth, 80
Transfer Act, 147
Transportation, 13-14, 119-21, 145, 251-52
"Trapper, Mad," story of, 195-98
Trappers. *See* Fur trade
Travel, cost of, 179
Treaty of Utrecht, 88
Tributary rivers, 5, 8, 38, 47, 117
Tripe de roche, 64
Tuktuk (Port Brabant) 154, 155, 162
Turnagain Point, 63, 70, 72, 78, 80, 81

U-235, 171, 183, 230-31
Union of Soviet Socialist Republics, relations with, 244-45
Union Strait, 71
United States, defense agreement with Canada, 247-51
Uranium, 4, 5, 6, 8, 159, 160, 171, 183, 201
production by LaBine mine, 227-36
search for new sources, 233-36
Utrecht, Treaty of, 88

Vancouver, 219, 220
Vermilion, Fort, 172
Victoria Island, 77, 191
Victoria Strait, 77, 84
Victory (ship), 72, 73
Vikings, search for the Northwest Passage, 56, 57

Walli, Emil, 234
War in the fur trade. *See* Fur trade
War, World II. *See* World War II
Watershed, 5, 6,
Waterways, Alberta, 5
Watson Lake, 212, 219, 220, 221, 222, 226
Way *into* North America, 244-45
Wedderburn, Fort, 114
Welcome (ship), 81
Wells, Norman. *See* Norman, Fort
Western Canada Airways, 173, 178, 189, 193
Whale Island, 45
Wheat lands, 5
Whitehorse, 212, 220, 221, 226
Whites, Indian contact with, 24-45
William, Fort, 109
Williams, William, 114
Williams Lake, 220
Winnipeg, 103
Winnipeg Lake, 107
Winter, length of, 24
Winter closing of the river. *See* Ice
Winter Lake, 63
Wollaston Land, 71
Wolseley, Colonel, 152
Women, Indian, status of, 25-26
World War II, 4, 12, 159-60, 208, 209-40, 243
Wrangel, Alaska, 219
Wrigley, 8, 132, 136, 156, 162
  landing field, 224

XY Company (Forsyth, Richardson and Company), 95, 97
  amalgamation with the North West Company, 98
  rivalry with the North West Company, 99-102

Yellowknife, 4, 5, 11, 13, 155, 158, 239
  gold rush, 175
  growth, 201-03
  landing field, 224
York boats, and era of, 86, 119-20, 135, 163
  production of, 137
York Factory, 105, 119, 120
York River, 107
Yukon, Fort, 131, 140-41
Yukon River, 46, 122, 140, 142
  discoverer of, 131
Yukon-Southern Air Transport, 212